# Modeling for Additive Manufacturing of Metals

PROCEEDINGS OF A WORKSHOP

Janki Patel, *Rapporteur*

Board on Mathematical Sciences and Analytics

National Materials and Manufacturing Board

Division on Engineering and Physical Sciences

*The National Academies of*
SCIENCES · ENGINEERING · MEDICINE

THE NATIONAL ACADEMIES PRESS
*Washington, DC*
**www.nap.edu**

THE NATIONAL ACADEMIES PRESS 500 Fifth Street, NW Washington, DC 20001

This activity was supported by the U.S. National Institute of Standards and Technology (Contract No. SB134117CQ0017), the Department of Energy, Los Alamos National Laboratory, and Sandia National Laboratories. Any opinions, findings, conclusions, or recommendations expressed in this publication do not necessarily reflect the views of any organization or agency that provided support for the project.

International Standard Book Number-13: 978-0-309-49420-5
International Standard Book Number-10: 0-309-49420-6
Digital Object Identifier: https://doi.org/10.17226/25481

Additional copies of this publication are available for sale from the National Academies Press, 500 Fifth Street, NW, Keck 360, Washington, DC 20001; (800) 624-6242 or (202) 334-3313; http://www.nap.edu.

Copyright 2019 by the National Academy of Sciences. All rights reserved.

Printed in the United States of America

Suggested citation: National Academies of Sciences, Engineering, and Medicine. 2019. *Data-Driven Modeling for Additive Manufacturing of Metals: Proceedings of a Workshop*. Washington, DC: The National Academies Press. https://doi.org/10.17226/25481.

*The National Academies of*
SCIENCES · ENGINEERING · MEDICINE

The **National Academy of Sciences** was established in 1863 by an Act of Congress, signed by President Lincoln, as a private, nongovernmental institution to advise the nation on issues related to science and technology. Members are elected by their peers for outstanding contributions to research. Dr. Marcia McNutt is president.

The **National Academy of Engineering** was established in 1964 under the charter of the National Academy of Sciences to bring the practices of engineering to advising the nation. Members are elected by their peers for extraordinary contributions to engineering. Dr. John L. Anderson is president.

The **National Academy of Medicine** (formerly the Institute of Medicine) was established in 1970 under the charter of the National Academy of Sciences to advise the nation on medical and health issues. Members are elected by their peers for distinguished contributions to medicine and health. Dr. Victor J. Dzau is president.

The three Academies work together as the **National Academies of Sciences, Engineering, and Medicine** to provide independent, objective analysis and advice to the nation and conduct other activities to solve complex problems and inform public policy decisions. The National Academies also encourage education and research, recognize outstanding contributions to knowledge, and increase public understanding in matters of science, engineering, and medicine.

Learn more about the National Academies of Sciences, Engineering, and Medicine at **www.nationalacademies.org**.

*The National Academies of*
## SCIENCES · ENGINEERING · MEDICINE

**Consensus Study Reports** published by the National Academies of Sciences, Engineering, and Medicine document the evidence-based consensus on the study's statement of task by an authoring committee of experts. Reports typically include findings, conclusions, and recommendations based on information gathered by the committee and the committee's deliberations. Each report has been subjected to a rigorous and independent peer-review process and it represents the position of the National Academies on the statement of task.

**Proceedings** published by the National Academies of Sciences, Engineering, and Medicine chronicle the presentations and discussions at a workshop, symposium, or other event convened by the National Academies. The statements and opinions contained in proceedings are those of the participants and are not endorsed by other participants, the planning committee, or the National Academies.

For information about other products and activities of the National Academies, please visit www.nationalacademies.org/about/whatwedo.

## PLANNING COMMITTEE ON THE WORKSHOP ON THE FRONTIERS OF MECHANISTIC DATA-DRIVEN MODELING FOR ADDITIVE MANUFACTURING

CAROLIN KÖRNER, Friedrich-Alexander Universität Erlangen-Nürnberg, *Co-Chair*
WING KAM LIU, Northwestern University, *Co-Chair*
TAHANY EL-WARDANY, United Technologies Research Center
ADE MAKINDE, General Electric Global Research Center
MUSTAFA MEGAHED, ESI Group
CELIA MERZBACHER, SRI International
NANCY REID, NAS,[1] University of Toronto
JENS TELGKAMP, Airbus Operations GmbH
KAREN E. WILLCOX, University of Texas, Austin

*Staff*

MICHELLE SCHWALBE, Director, Board on Mathematical Sciences and Analytics, *Workshop Co-Director*
ERIK SVEDBERG, Senior Program Officer, National Materials and Manufacturing Board, *Workshop Co-Director*
BETH DOLAN, Financial Manager
JANKI PATEL, Research Associate

---

[1] Member, National Academy of Sciences.

# BOARD ON MATHEMATICAL SCIENCES AND ANALYTICS

MARK L. GREEN, University of California, Los Angeles, *Chair*
HÉLÈNE BARCELO, Mathematical Sciences Research Institute
JOHN R. BIRGE, NAE,[1] University of Chicago
W. PETER CHERRY, NAE, Independent Consultant
DAVID S.C. CHU, Institute for Defense Analyses
RONALD R. COIFMAN, NAS,[2] Yale University
JAMES (JIM) CURRY, University of Colorado, Boulder
SHAWNDRA HILL, Microsoft Research
LYDIA KAVRAKI, NAM,[3] Rice University
TAMARA KOLDA, Sandia National Laboratories
JOSEPH A. LANGSAM, University of Maryland
DAVID MAIER, Portland State University
LOIS CURFMAN McINNES, Argonne National Laboratory
JILL C. PIPHER, Brown University
ELIZABETH A. THOMPSON, NAS, University of Washington
CLAIRE J. TOMLIN, NAE, University of California, Berkeley
LANCE A. WALLER, Emory University
KAREN E. WILLCOX, University of Texas, Austin

*Staff*

MICHELLE SCHWALBE, Director
TYLER KLOEFKORN, Program Officer
LINDA CASOLA, Associate Program Officer
ADRIANNA HARGROVE, Financial Manager
SELAM ARAIA, Program Assistant

---

[1] Member, National Academy of Engineering.
[2] Member, National Academy of Sciences.
[3] Member, National Academy of Medicine.

# NATIONAL MATERIALS AND MANUFACTURING BOARD

BEN WANG, Georgia Institute of Technology, *Chair*
THERESA KOTANCHEK, Evolved Analytics, LLC, *Vice Chair*
RODNEY C. ADKINS, NAE,[1] IBM Corporate Strategy (Retired)
CRAIG ARNOLD, Princeton University
JIM C.I. CHANG, National Cheng Kung University, Tainan, Taiwan
THOMAS M. DONNELLAN, The Pennsylvania State University
STEPHEN R. FORREST, NAS[2]/NAE, University of Michigan
ERICA R.H. FUCHS, Carnegie Mellon University
DAVID C. LARBALESTIER, NAE, Florida State University
MICHAEL MAHER, Maher and Associates, LLC
ROBERT D. MILLER, NAE, IBM Almaden Research Center
EDWARD MORRIS, Consequence Consulting, LLC
NICHOLAS A. PEPPAS, NAE/NAM,[3] University of Texas, Austin
TRESA M. POLLOCK, NAE, University of California, Santa Barbara
GREGORY TASSEY, University of Washington
HAYDN WADLEY, University of Virginia
STEVEN J. ZINKLE, NAE, University of Tennessee, Knoxville

*Staff*

JAMES LANCASTER, Director
ERIK SVEDBERG, Senior Program Officer
NEERAJ P. GORKHALY, Associate Program Officer
HEATHER LOZOWSKI, Financial Associate
BETH DOLAN, Financial Associate
AMISHA JINANDRA, Research Associate
JOSEPH PALMER, Senior Project Assistant

---

[1] Member, National Academy of Engineering.
[2] Member, National Academy of Sciences.
[3] Member, National Academy of Medicine.

# Acknowledgment of Reviewers

This Proceedings of a Workshop was reviewed in draft form by individuals chosen for their diverse perspectives and technical expertise. The purpose of this independent review is to provide candid and critical comments that will assist the National Academies of Sciences, Engineering, and Medicine in making each published proceedings as sound as possible and to ensure that it meets the institutional standards for quality, objectivity, evidence, and responsiveness to the charge. The review comments and draft manuscript remain confidential to protect the integrity of the process.

We thank the following individuals for their review of this proceedings:

Bianca Maria Colosimo, Politecnico di Milano;
Joseph M. DeSimone, NAS[1]/NAE[2]/NAM,[3] University of North Carolina, Chapel Hill; and
John Turner, Oak Ridge National Laboratory.

Although the reviewers listed above provided many constructive comments and suggestions, they were not asked to endorse the content of the proceedings nor did they see the final draft before its release. The

---

[1] Member, National Academy of Sciences.
[2] Member, National Academy of Engineering.
[3] Member, National Academy of Medicine.

review of this proceedings was overseen by Mark C. Hersam, Northwestern University. He was responsible for making certain that an independent examination of this proceedings was carried out in accordance with the standards of the National Academies and that all review comments were carefully considered. Responsibility for the final content rests entirely with the rapporteur and the National Academies.

# Contents

1 INTRODUCTION 1
  Organization of This Proceedings, 2

2 PROCESS MONITORING AND CONTROL 3
  Measurements and Modeling for Process Monitoring
     and Control, 4
  Measurement Science for Process Monitoring and Control, 9
  Simulations: A Chance for Knowledge-Based Improvement
     of Additive Manufacturing, 11
  Discussion, 13
  References, 15

3 MICROSTRUCTURE EVOLUTION, ALLOY DESIGN,
  AND PART SUITABILITY 19
  Measurements for Additive Manufacturing of Metals, 19
  Predicting Material State and Performance of
     Additively Manufactured Parts, 22
  Discussion, 24
  References, 27

4   PROCESS AND MACHINE DESIGN                                           28
    Modeling Phases of Process and Machine Design, 29
    Current State of Commercial Powder-Bed Additive
        Machines—AM Machine Design Issues Impacting
        Build-to-Build and Part-to-Part Variability, 33
    Modeling Challenges and Opportunities at the Part Level, 35
    Discussion, 39
    References, 39

5   PRODUCT AND PROCESS QUALIFICATION
    AND CERTIFICATION                                                    41
    Process Qualification and Technological Validation,
        from Casting to Additive, 41
    Modeling and Simulation, 43
    Discussion, 44
    Reference, 46

6   SUMMARY OF CHALLENGES FROM SUBGROUP
    DISCUSSIONS AND PARTICIPANT COMMENTS                                 47
    Measurements and Modeling for Process Monitoring
        and Control, 48
    Developing Models to Represent Microstructure
        Evolution, Alloy Design, and Part Suitability, 50
    Modeling Aspects of Process and Machine Design, 52
    Accelerating Product and Process Qualification
        and Certification, 54
    Individual Response Results, 56
    References, 56

APPENDIXES
A   Registered Workshop Participants                                     59
B   Workshop Agenda                                                      61
C   Workshop Statement of Task                                           66

# 1

# Introduction

Additive manufacturing (AM), the process in which a three-dimensional object is built by adding subsequent layers of materials, has the potential to revolutionize how mechanical parts are created, tested, and certified. AM enables novel material compositions and shapes, often without the need for specialized tooling. However, successful real-time AM design requires the integration of complex systems and often necessitates expertise across domains.

The complex design and processing systems that enable AM start with computer models. Since AM processes can be difficult to measure experimentally and empirical models for AM can be expensive to create, advanced fundamental models (including mechanistic data-driven reduced-order models and other validated theoretical and computational models) can be used to better understand underlying physical mechanisms. Simulation-based design approaches, such as those applied in engineering product design and material design, have the potential to improve AM predictive modeling capabilities, particularly when combined with existing knowledge of the underlying mechanics. These predictive models have the potential to reduce the cost of and time for concept-to-final-product development and can be used to supplement experimental tests.

On October 24–26, 2018, the National Academies of Sciences, Engineering, and Medicine organized a workshop of experts from various communities within the United States and the European Union to discuss the frontiers of mechanistic data-driven modeling for AM of metals. The planning committee (shown on page v) helped to identify the workshop topics, invite speakers, and plan the agenda. The workshop was held at the Neue Materialien Fürth GmbH building of the Friedrich-Alexander-Universität Erlangen-Nürnberg in Fürth, Germany. This workshop was sponsored by the U.S. Department of Energy, the U.S. National Institute of Standards and Technology, Sandia National Laboratories, and Los Alamos National Laboratory.

Wing Kam Liu (Northwestern University), the chair of the planning committee, opened the workshop by discussing its four main topics:

- Measuring and modeling process monitoring and control;
- Developing models to represent microstructure evolution, alloy design, and part suitability;
- Modeling phases of process and machine design; and
- Accelerating product and process qualification and certification.

The first 2 days of the workshop focused on presentations and panel discussions relating to the workshop themes. The third day centered on breakout groups that discussed some of the short-, intermediate-, and long-term challenges in AM.

This proceedings summarizes the presentations and discussions that took place during the workshop. The viewpoints expressed in this proceedings are those of individual workshop participants and do not necessarily represent the views of all workshop participants, the planning committee, or the National Academies of Sciences, Engineering, and Medicine.

## ORGANIZATION OF THIS PROCEEDINGS

The following chapters in this proceedings summarize the workshop's presentations and discussions. Chapter 2 describes the measurements and modeling for process monitoring control in AM. Chapter 3 provides an overview of developing models to represent microstructure evolution, alloy design, and part suitability. Chapter 4 focuses on modeling aspects of process and machine design. Chapter 5 discusses opportunities to accelerate product and process qualification and certification. Chapter 6 summarizes challenges raised during subgroup discussions and by individual participants. A list of registered workshop participants appears in Appendix A, and Appendix B includes the 3-day workshop agenda.

# 2

# Process Monitoring and Control

The first workshop session provided an overview of measurements and modeling for process monitoring and control in additive manufacturing (AM). Speakers described systems measured in situ and in real time as well as challenges of each resolution and process signature. Bianca Maria Colosimo (Politecnico di Milano), Jarred Heigel (National Institute of Standards and Technology), Marvin Siewert (University of Bremen), Kilian Wasmer (Empa), Ben Dutton (Manufacturing Technology Centre), and Amit Surana (United Technologies Research Center) each discussed research, challenges, and future directions relating to the following questions:

- How can systems be measured in real time?
- What AM measurements enable uncertainty quantification?
- How can the precision of a measurement be certified?
- How can measured data be employed to understand the full state of a system?
- What mathematical and statistical methods could be applied to AM? How can resources from other disciplines be integrated?
- What can be measured in situ and in line? What are the main challenges of coaxial and off-axis sensing in terms of accuracy, frequency, and spatial and temporal resolution?
- What is the correlation between process signature and product defects? How does the probability of detecting flaws connect with the qualification of an additively manufactured item?

- How can models and solutions be used to transfer knowledge from machine to machine and from laboratory to laboratory? How does this change depending on the material and geometry selected to make a part?
- What are the impacts of false positives and false negatives? What are the economic advantages of in-situ monitoring?
- What are the challenges of moving from monitoring to feedback control?

## MEASUREMENTS AND MODELING FOR PROCESS MONITORING AND CONTROL

*Bianca Maria Colosimo, Politecnico di Milano*

Colosimo described Politecnico di Milano's AddMe.Lab, a laboratory combining industrial machines and novel prototypes for AM processes such as selective laser melting, electron beam melting, directed energy deposition with powder and wire feedstocks, and binder jetting. She explained that in-situ monitoring can help reduce major industrial barriers for metal AM technologies, such as process instability, lack of repeatability, and defect rates (Mani et al., 2017; Tapia and Elwany, 2014; Everton et al., 2016; Spears and Gold, 2016; Grasso and Colosimo, 2017).

Defects in AM products originate in a variety of ways, including the equipment, process, design choices, and feedstock material. Colosimo shared several references for defect sources, as shown in Table 2.1. The process signature, which represents the manufacturing process through which data are collected from control systems and sensors, can give insights into approaches to control the quality of the final product. Ideally, in-situ monitoring could identify defects in real time and correct the process accordingly.

Colosimo provided examples of different levels of in-situ monitoring.

- Level 0: Using the existing signals (without additional sensors) to appropriately analyze all of the available information via statistical machine learning in order to predict defect onset from monitoring and fusing signal data (Grasso and Colosimo, 2017).
- Level 1: Monitoring the powder bed to assess uniformity of the powder coverage, the geometry, and possibly the temperature distribution of the melted layer. These assessments can be done using high-resolution images in the visible and infrared bands. At this level, it is possible to detect delamination defects as well as geometrical deviation between the actual and the nominal shape

TABLE 2.1 Defect Sources and Categories by Publication

| Sources of defects | | Categories of defects | | | | | |
|---|---|---|---|---|---|---|---|
| | | Porosity | Balling | Geometric defects | Surface defects | Residual stresses, cracks, and delamination | Microstructural inhomogeneity and impurity |
| Equipment | Beam scanning/ deflection | Foster et al., 2015 | | Moylan et al., 2014; Foster et al., 2015 | | | |
| | Build chamber environment | Ferrar et al., 2012; Spears and Gold, 2016 | Li et al., 2012 | | | Edwards et al., 2013; Chlebus et al., 2011; Buchbinder et al., 2014; Kempen et al., 2013 | Spears and Gold, 2016 |
| | Powder handling and deposition | Foster et al., 2015 | | Foster et al., 2015; Kleszczynski et al., 2012 | Foster et al., 2015; Kleszczynski et al., 2012 | | Foster et al., 2015 |
| | Baseplate | | | Prabhakar et al., 2015 | | Prabhakar et al., 2015 | |

*continued*

**TABLE 2.1** Continued

| Sources of defects | | Categories of defects | | | | | |
|---|---|---|---|---|---|---|---|
| | | Porosity | Balling | Geometric defects | Surface defects | Residual stresses, cracks, and delamination | Microstructural inhomogeneity and impurity |
| Process | Parameters and scan strategy | Matthews et al., 2016; Yasa et al., 2009; Attar, 2011; Gong et al., 2013; Read et al., 2015; Kruth et al., 2004; Weingarten et al., 2015; Thijs et al., 2010; Scharowsky et al., 2015; Puebla et al., 2012; Tammas-Williams et al., 2015; Biamino et al., 2011; Zeng, 2015 | Li et al., 2012; Kruth et al., 2004; Tolochko et al., 2004; Zhou et al., 2015; Attar, 2011; Gong et al., 2013 | Yasa et al., 2009; Mousa, 2016; Kleszczynski et al., 2012; Thomas, 2009 | Li et al., 2012; Kruth et al., 2004; Matthews et al., 2016; Attar, 2011; Gong et al., 2013; Zaeh and Kanhert, 2009; Delgado et al., 2012 | Mercelis and Kruth, 2006; Parry et al., 2016; Cheng et al., 2016; Van Belle et al., 2013; Casavola et al., 2008; Zäh and Lutzmann, 2010; Zaeh and Branner, 2010; Kempen et al., 2013; Kruth et al., 2004; Carter et al., 2014 | Carter et al., 2014; Arisoy et al., 2017; Niu and Chang, 1999; Huang et al., 2016; Thijs et al., 2010; Scharowsky et al., 2015; Puebla et al., 2012; Biamino et al., 2011 |
| | Byproducts and material ejections | Liu et al., 2015; Khairallah et al., 2016 | | | | | Liu et al., 2015; Khairallah et al., 2016 |

| | | | | | | |
|---|---|---|---|---|---|---|
| Design choices | Supports | | | Foster et al., 2015; Kleszczynski et al., 2012; Zeng, 2015 | Foster et al., 2015; Kleszczynski et al., 2012; Zeng, 2015 | Foster et al., 2015; Kleszczynski et al., 2012; Zeng, 2015 |
| | Orientation | Li et al., 2012; Strano et al., 2013 | Delgado et al., 2012 | Delgado et al., 2012; Fox et al., 2016; Strano et al., 2013 | | Meier and Haberland, 2008 |
| | | | Das, 2003 | Seyda et al., 2012 | | |
| Feedstock material (powder) | Liu et al., 2015; Van Elsen, 2007; Das, 2003 | | | | | Das, 2003; Niu and Chang, 1999; Huang et al., 2016 |

SOURCE: M. Grasso and B.M. Colosimo, 2017, Process defects and in-situ monitoring methods in metal powder-bed fusion: A review, *Measurement Science and Technology* 28(4):1–25, 10.1088/1361-6501/aa5c4f. © IOP Publishing. Reproduced with permission. All rights reserved.

printed at each layer (Tapia and Elwany, 2014; Mani et al., 2017; Grasso and Colosimo, 2017; Everton et al., 2016).
- Level 2: Monitoring the printed layer, using high-speed videos in the visible or infrared ranges. Hot and cold spots (i.e., areas that remain hot or cold for a long period of time and can cause geometrical or volumetric defects due to over-melting or under-melting) can be detected in the thermal signature. Infrared video cameras can aid in computing the spatial gradient and temporal gradient, which can be used to predict the final microstructure (Land et al., 2015; Krauss et al., 2014; Caltanissetta et al., 2018; Arnold et al., 2018; Trushnikov et al., 2016; Grasso and Colosimo, 2016; Colosimo and Grasso, 2018; Brumana et al., 2018).
- Level 3: Monitoring the AM track to assess the spatter signature, the plume, the shape of the track, and the cooling rate left by the beam behind it. The spatter signature can relate to the expected porosity, and an excessive plume can lead to job failure (Repossini et al., 2017; Ly et al., 2017).
- Level 4: Monitoring the melt-pool size, shape, and temperature. Since the laser directly impacts the melt pool, feedback control could be implemented to keep the melt-pool signature stable by varying the laser power and/or speed (Doubenskaia et al., 2012; Berumen et al., 2010; Kruth et al., 2007).

Levels 1 through 3 are considered "off-axis monitoring" because they need sensors that are not placed coaxially with the laser beam.

Colosimo emphasized that in-situ sensing can improve understanding of the AM process, allow for calibration of the AM process simulations, increase part quality (e.g., by detecting, preventing, or even compensating for defects), and support process qualification. Some pending issues, however, include correlating the process signature with product quality and modeling defects appropriately. She also outlined key sensing questions: How should the appropriate sensors and their spatial and temporal resolutions be chosen? How could in-situ sensing accuracy be certified? What methods and tools should be used for multisensor data fusion?

A goal is to move from "sensorized" machines that collect data to "intelligent" AM systems that use data to make decisions. This transition requires a combination of statistical methods to visualize effectiveness and efficiency. Colosimo stressed that multidisciplinary research is needed to enable new ideas in in-situ sensing, monitoring, and control.

## MEASUREMENT SCIENCE FOR PROCESS MONITORING AND CONTROL

*Jarred Heigel, National Institute of Standards and Technology*

Heigel described the eight project areas of the Measurement Science for Additive Manufacturing program within the Engineering Laboratory at the National Institute of Standards and Technology (NIST):

1. Precursor material qualification,
2. AM machine and process qualification,
3. AM part qualification,
4. Metrology for multiphysics AM model validation,
5. Metrology for real-time monitoring of AM,
6. Machine and process control methods for AM,
7. Data-driven decision support for AM, and
8. Data integration and management for AM.

The primary objective of this program is to "develop and deploy measurement science that will enable rapid design-to-product transformation through advances in material characterization; in-process process sensing, monitoring, and model-based optimal control; performance qualification of materials, machines, processes, and parts; and end-to-end digital implementation and integration of AM processes and systems" (NIST, 2019). Heigel's presentation focused on measurements and sensors used for real-time monitoring, challenges of real-time monitoring and control, and the path forward.

Real-time monitoring, Heigel stated, includes any sensor measurements that are continuously recorded during the AM process and used to ensure that the machine and process are performing as expected.[1] Common optical sensors include high-speed cameras, pyrometers, in-line cameras, and in-line photodetectors. These optical sensors can provide great insight into each layer but are limited to observing only the surfaces. Ultrasonic sensors can be used to detect subsurface defects by sending ultrasonic waves through the part, and acoustic sensors can detect melt-pool quality and part failure by monitoring the acoustic emissions from the melt pool and cracks.

Heigel explained that real-time monitoring enables both statistical process control and feedback control. Statistical process control involves comparing the data from a new build with historical data of other builds

---

[1] In the context of this presentation, layer-wise imaging or intermittent measurements are not considered real-time monitoring.

to determine whether the process is performing within an acceptable range. It also involves collecting data from process signatures and comparing them with control limits, which are calculated for the expected measurements of the process output. In contrast, feedback control relies on real-time monitoring and high-rate continuous measurement analysis that can then be used to modify the process.

The industry is striving for rapid processes and rapid certification, with the help of real-time monitoring and associated control. However, Heigel explained that the largest current barriers for industry are high capital costs, a lack of robust correlations, and difficulty interpreting what is being measured. Different monitoring approaches are being deployed to balance cost and speed constraints. Some AM machines are enabling layer-wise imaging and melt-pool monitoring. Coaxial photodetectors enable low-cost monitoring at sufficiently high speeds (compared to high-speed imaging) but lack fidelity to interpret processing quality. For directed energy deposition systems, coaxial melt-pool imaging is currently being used for real-time monitoring and control and feedback control because the process dynamics are comparatively slower.

Heigel elaborated on some challenges for real-time monitoring. The first challenge mentioned was measurement fidelity, which involves the trade-off between high spatial resolution and high temporal resolution. Thermal cameras can provide high spatial resolution but are temporally limited to $10^3$ Hz. Photodetectors can provide higher temporal resolution but cannot directly determine dynamic size variations in the melt pool. Another challenge for real-time monitoring is correlating the sensor data with the physics underlying the AM process. Better understanding the physics helps to inform decisions about what types of sensors to use, how to interpret the measurements, how to calibrate those measurements, what control algorithms to use, and how to prioritize research and development.

However, real-time monitoring and feedback control cannot correct flawed designs or processes. Heigel explained that the process must be improved to minimize variability, and the build strategies must be designed to optimize the process. The importance of modeling and validation efforts toward achieving this goal cannot be overstated. In addition, an improved understanding of the relationship between defects and real-time monitoring signals must be developed. This requires improvements to the post-process detection of defects and consideration of how the real-time data are processed and stored. Finally, metrology improvements, such as better calibration of the sensors, will play an important role in allowing data acquired across machines to be compared.

During the question and answer portion of this presentation, a participant asked Heigel what to do if a defect is detected during the monitoring process. Should the part be discarded or is there a way to fix it? Heigel responded that, first, one has to identify the type of defect accurately and determine whether it is fixable. For example, it may be possible to re-scan a pore relatively close to the surface and to release the associated void, but a part with a crack may have to be discarded. Real-time monitoring can help determine whether a defective part should be discarded, corrected, or ignored.

Workshop co-chair Wing Kam Liu (Northwestern University) asked about the challenges of powder-bed technologies versus multiple-head machines. Heigel said that NIST focuses on powder-bed systems due to limited resources. While both are being used in industry, the focus tends to be on powder-bed technologies. Heigel emphasized that both methods have important considerations and would benefit from additional research. He noted that there are some challenges in powder-bed versus directed energy deposition. For example, ultrasonic measurements tend to be transferrable, but differences in process speed can create different sized melt pools and cause a different formation. Also, differences in the plumes and powder delivery result in different types of problems in powder-bed and multiple-head technologies. Lessons learned from measurement science about different optimal sensors, ultrasonics, and acoustics could be applied to both technologies.

## SIMULATIONS: A CHANCE FOR KNOWLEDGE-BASED IMPROVEMENT OF ADDITIVE MANUFACTURING

*Marvin Siewert, University of Bremen*

Siewert began by explaining four competences in AM: (1) part design (e.g., topology optimization, residual stress and distortion, compensation of distortion), (2) pre-processing (e.g., part orientation, support structures, nesting of parts), (3) process (e.g., scan strategies, thermal management, microstructure properties), and (4) post treatment (e.g., hot isostatic pressing, milling, heat treatment). He provided several examples of how these competences work together in practice.

The first example was a simulation of residual stress and distortion. The classical thermomechanical approach calculates the temperature field using the initial condition and suitable boundary conditions. Next, thermal strains and force equilibrium are calculated at several time steps. Siewert noted that while this approach can be informative, it can also be difficult to calibrate and validate as well as time-consuming and cost-intensive to run. In contrast, the mechanical process equivalent method requires inserting the inherent strains of the union of multiple layers as loads into a mechanical calculation. This approach can be calibrated more easily, even for large parts, and is implemented within software at the Integrated Status and Effectiveness Monitoring Program (ISEMP). The Additive Works GmbH is a spin-off company of ISEMP and provides (among other things) simulations based on the mechanical process equivalent method in its software Amphyon.[2] Some goals and applications for this simulation are fast computation of residual stress and distortion, fast estimation, identification of critical areas, adaptation of the design, and simulation-based adaptation of support structures. Ade Makinde (General Electric Global Research Center) wondered about the accuracy of the mechanical process equivalent method. Siewert noted that shrinkage from every new layer calibrates well and does not require data to be uploaded, which enables faster computations.

Siewert's second example described mesoscopic and macroscopic simulations of a temperature field. On the mesoscopic scale, a Goldak heat source was used to analyze the melt-pool size and shape. This type of analysis can be used to calibrate heat sources by comparing microsections with simulated melt-pool shapes, to explore the influence of local geometry on the melt-pool size (e.g., overhanging regions with different angles), and to estimate cooling rates. On the macroscopic scale, energy input is realized by element activation at a certain temperature. This type of analysis can be used, for example, to understand the influence of different hatch orders (i.e., the order in which material is filled within the boundaries of AM parts) and to identify critical areas (e.g., hot spots). The models on both scales use fast finite-difference method/finite-element method calculations and are currently undergoing experimental validation.

Siewert explained that the vision of predicting and controlling all parameters in the entire AM process requires broad and deep thinking. Since the quality and reliability of the produced part is influenced by the whole process chain, every step needs to be understood as well as possible. Simulation methods and algorithms are needed to understand the process and measured data, to predict critical situations, to adapt to and overcome these situations, and to optimize the process. He emphasized that data to validate and calibrate methods as well as improved mechanisms to get adapted parameters into the process are critical to realize this vision.

In response to a question about employing measured data to understand the full state of a system, Siewert said that one could use measured data to calibrate and validate a simulation model. A reliable simulation can give a deeper understanding of the process within the system.

---

[2] To learn more about Amphyon, see https://web.altair.com/2017-introduction-to-additive-works, accessed October 26, 2018.

Another participant asked how the accuracy and precision of a measurement can be certified. Siewert explained that although applications vary, different measurement techniques and the calibrated simulations can be combined to improve the reliability of measurements.

## DISCUSSION

Heigel, Wasmer, Siewert, Dutton, and Surana participated in a panel discussion moderated by Colosimo. Colosimo asked each panelist to comment on sensor issues. Wasmer said that the cost of sensors will always be a consideration as sensors may be too expensive for some low-cost applications. He also noted the value of a central resource for results from various types of sensors. Researchers have to understand as much as they can about both the process and the limits of the sensors in order to minimize error; often this can be done by measuring the piece directly, independent from the process parameters. A participant from industry asked which sensor data are most helpful. Surana responded that sensing could be used in many different ways, including off-line model validation and in-line detection of failure. However, the sensing process also depends on the scale being modeled and the techniques being used.

A participant asked the panelists to comment on the repeatability of sensor data, how consistent the sensors are across machines and manufacturers, and standards for these sensors. Heigel mentioned that NIST has been working toward understanding both sensor variability and machine/process variability and that a lack of standards or best practices for calibration is a barrier. NIST has been conducting an interlaboratory study for the past few years to investigate powder-bed fusion variability, and some irregularities have been observed. NIST is also researching the physics behind sensor measurement and developing calibration procedures. Dutton added that it is important to develop these tools to enable the sensors to scale up into other ranges. He added that a structural model of the part capabilities as well as the type(s) of defects and sizes that the part can handle would be helpful in establishing quality requirements for a part.

In response to a question about the effect of sensor distribution, Dutton mentioned that most current sensing methods are only looking at the top surface and can miss deeper defects. Other methods not yet applied, such as laser ultrasound, could cover both surfaces and material within about 2 mm of the surface. Improved sensing during the layer-by-layer AM build process may enable more thorough defect detection. Colosimo added that it is difficult to learn across machines because sensor integration tends to be manufacturer-specific. She also mentioned that multiple sensors could be used to increase the robustness of results and detect when problems with a specific sensor occur. Another audience member asked whether NIST is considering any compression or filtering techniques to reduce storage demands for large data sets. Heigel responded that although NIST is not specifically looking into this, data storage is an important consideration. Colosimo encouraged widespread data sharing to enable faster progress. Heigel commented that although data sharing can help identify mistakes, open communication is essential; data can be easily misinterpreted, and challenges exist with repurposing data to new applications. Surana suggested that some data may be prioritized for storage along with summaries of where the supplemental information can be found.

In response to a question about the importance of process parameters, Wasmer emphasized the value of determining the exact moment something happens so that that moment can be explored using other techniques, such as machine learning. Surana agreed that this is an important opportunity. Another participant asked the panelists to comment on the challenges associated with part inspection. Colosimo responded that in-situ monitoring allows some visibility into the process during the build but is not as helpful when defects depend on the post-processing steps (e.g., thermal treatment and finishing).

Liu asked the panelists for their thoughts on short-term, intermediate, and long-term goals in AM. The panelists suggested the following areas for improvement:

- *Short-term goals*
  — Improving imaging capabilities (Colosimo);
  — Clarifying what to monitor and when (Dutton);
  — Setting expectations for assessing what can and cannot be done (Heigel); and
  — Establishing calibration procedures (Heigel).
- *Intermediate-term goals*
  — Facilitating real-time feedback control (Colosimo);
  — Improving the use of models and statistical analyses to determine the optimal level of feedback, taking into consideration the design and purpose of the AM part (Dutton);
  — Improving modeling capabilities to predict and design the process (Heigel); and
  — Advancing fast computations (Surana).
- *Long-term goals*
  — Improving the fundamental understanding of the processes, especially for varying shapes and materials (Colosimo);
  — Designing processes to be consistent across machines and sensors (Heigel); and

— Encouraging lifelong learning with respect to new parts, processes, and data management (Surana).

## REFERENCES

Arisoy, Y.M., L.E. Criales, T. Özel, B. Lane, and S. Moylan. 2017. Influence of scan strategy and process parameters on microstructure and its optimization in additively manufactured nickel alloy 625 via laser powder bed fusion. *The International Journal of Advanced Manufacturing Technology* 90(5–8):1393–1417.

Arnold, C., C.R. Pobel, F. Osmanlic, and C. Körner. 2018. Layerwise monitoring of electron beam melting via backscatter electron detection. *Rapid Prototyping Journal* 24(8):1401–1406.

Attar, E. 2011. Simulation of selective electron beam melting processes [Dr.-Ing.]. Friedrich-Alexander Universität Erlangen-Nürnberg.

Berumen, S., F. Bechmann, S. Lindner, J.-P. Kruth, and T. Craeghs. 2010. Quality control of laser- and powder bed-based additive manufacturing (AM) technologies. *Physics Procedia* 5(B):617–622.

Biamino, S., A. Penna, U. Ackelid, S. Sabbadini, O. Tassa, P. Fino, M. Pavese, P. Gennaro, and C. Badini. 2011. Electron beam melting of Ti-48Al-2Cr- 2Nb alloy: Microstructure and mechanical properties investigation. *Intermetallics* 19(6):776–781.

Brumana, A., G. Fazzini, M. Grasso, and B.M. Colosimo. 2018. In-situ thermographic measurement during additive manufacturing selective laser melting [Master thesis]. Politecnico di Milano, Milan, Italy.

Buchbinder, D., W. Meiners, N. Pirch, K. Wissenbach, and J. Schrage. 2014. Investigation on reducing distortion by preheating during manufacture of aluminum components using selective laser melting. *Journal of Laser Applications* 26(1):012004.

Caltanissetta, F., M. Grasso, S. Petrò, and B. Colosimo. 2018. Characterization of in-situ measurements based on layerwise imaging in laser powder bed fusion. *Additive Manufacturing* 24:183–199.

Carter, L.N., C. Martin, P.J. Withers, and M.M. Attallah. 2014. The influence of the laser scan strategy on grain structure and cracking behaviour in SLM powder-bed fabricated nickel super alloy. *Journal of Alloys and Compounds* 615:338–347.

Casavola, C., S.L. Campanelli, and C. Pappalettere. 2008. Experimental analysis of residual stresses in the selective laser melting process. In *Proceedings of the XIth International Congress and Exposition*. Society for Experimental Mechanics, Orlando, Fla., June 2–5. https://www.academia.edu/21603785/Experimental_analysis_of_residual_stresses_in_the_selective_laser_melting_process.

Cheng, B., S. Shrestha, and K. Chou. 2016. Stress and deformation evaluations of scanning strategy effect in selective laser melting. *Additive Manufacturing* 12(B):240–251.

Chlebus, E., B. Kuźnicka, and T. Kurzynowski. 2011. Microstructure and mechanical behaviour of Ti—6Al—7Nb alloy produced by selective laser melting. *Materials Characterization* 62(5):488–495.

Colosimo, B.M., and M. Grasso. 2018. Spatially weighted PCA for monitoring video image data with application to additive manufacturing. *Journal of Quality Technology* 50(4):391–417.

Das, S. 2003. Physical aspects of process control in selective laser sintering of metals. *Advanced Engineering Materials* 5(10):701–711.

Delgado, J., J. Ciurana, and C.A. Rodríguez. 2012. Influence of process parameters on part quality and mechanical properties for DMLS and SLM with iron-based materials. *The International Journal of Advanced Manufacturing Technology* 60(5-8):601–610.

Doubenskaia, M., M. Pavlov, S. Grigoriev, E. Tikhonova, and I. Smurov. 2012. Comprehensive optical monitoring of selective laser melting. *Journal of Laser Micro/Nanoengineering* 7(3):236–243.

Edwards, P., A. O'Conner, and M. Ramulu. 2013. Electron beam additive manufacturing of titanium components: Properties and performance. *Journal of Manufacturing Science and Engineering* 135(6):1–7.

Everton, S., M. Hirsch, P. Stavroulakis, R. Leach, and A. Clare. 2016. Review of in-situ process monitoring and in-situ metrology for metal additive manufacturing. *Materials and Design* 95:431–445.

Ferrar, B., L. Mullen, E. Jones, R. Stamp, and C.J. Sutcliffe. 2012. Gas flow effects on selective laser melting (SLM) manufacturing performance. *Journal of Materials Processing Technology* 212(2):355–364.

Foster, B.K., E.W. Reutzel, A.R. Nassar, C.J. Dickman, and B.T. Hall. 2015. A brief survey of sensing for metal-based powder bed fusion additive manufacturing. In *Proceedings of SPIE—The International Society for Optical Engineering* 9489: 94890B. Dimensional Optical Metrology and Inspection for Practical Applications IV, Baltimore, Md., April 20–21.

Fox, J.C., S.P. Moylan, and B.M. Lane. 2016. Effect of process parameters on the surface roughness of overhanging structures in laser powder-bed fusion additive manufacturing. *Procedia CIRP* 45:131–134.

Gong, H., K. Rafi, N.V. Karthik, T. Starr, and B. Stucker. 2013. Defect morphology in Ti-6Al-4V parts fabricated by selective laser melting and electron beam melting. Pp. 440–453 in *Proceedings of the Solid Freeform Fabrication Symposium*. 24th Annual International Solid Freeform Fabrication Symposium, Austin, Tex.

Grasso, M., and B.M. Colosimo. 2016. An automated approach to enhance multi-scale signal monitoring of manufacturing processes. *Journal of Manufacturing Science and Engineering* 138(5):051003–051003-16.

Grasso, M., and B.M. Colosimo. 2017. Process defects and in-situ monitoring methods in metal powder bed fusion: A review. *Measurement Science and Technology* 28(4):1–25.

Huang, Q., N. Hu, X. Yang, R. Zhang, and Q. Feng. 2016. Microstructure and inclusion of Ti-Al-4V fabricated by selective laser melting. *Frontiers of Materials Science* 10(4):428–431.

Kempen, K., L. Thijs, B. Vrancken, S. Buls, J. Van Humbeeck, and J.-P. Kruth. 2013. Producing crack-free, high density M2 HSS parts by selective laser melting: Pre-heating the baseplate. Pp. 131–139 in *2013 Proceedings of the Solid Freeform Fabrication Symposium*. 24th Annual International Solid Freeform Fabrication Symposium, Austin, Tex.

Khairallah, S.A., A.T. Anderson, A. Rubenchik, and W.E. King. 2016. Laser powder-bed fusion additive manufacturing: Physics of complex melt flow and formation mechanisms of pores, spatter, and denudation zones. *Acta Materialia* 108:36–45.

Kleszczynski, S., J. zur Jacobsmühlen, J.T. Sehrt, and G. Witt. 2012. Error detection in laser beam melting systems by high resolution imaging. Pp. 975–987 in *2012 Proceedings of the Solid Freeform Fabrication Symposium*. 24th Annual International Solid Freeform Fabrication Symposium, Austin, Tex.

Krauss, H., T. Zeugner, and M.F. Zaeh. 2014. Layerwise monitoring of the selective laser melting process by thermography. *Physics Procedia* 56:64–71.

Kruth, J.P., L. Froyen, J. Van Vaerenbergh, P. Mercelis, M. Rombouts, and B. Lauwers. 2004. Selective laser melting on iron-based powder. *Journal of Materials Processing Technology* 149:616–622.

Kruth, J.-P., G. Levy, F. Klocke, and T.H.C. Childs. 2007. Consolidation phenomena in laser and powder-bed based layered manufacturing. *CIRP Annals—Manufacturing Technology* 56(2):730–759.

Land, W.S., B. Zhang, J. Ziegert, and A. Davies. 2015. In-situ metrology system for laser powder-bed fusion additive process. *Procedia Manufacturing* 1:393–403.

Li, R., J. Liu, Y. Shi, L. Wang, and W. Jiang. 2012. Balling behavior of stainless steel and nickel powder during selective laser melting process. *The International Journal of Advanced Manufacturing Technology* 59(9–12):1025–1035.

Liu, Y., Y. Yang, S. Mai, D. Wang, and C. Song. 2015. Investigation into spatter behavior during selective laser melting of AISI 316L stainless steel powder. *Materials and Design* 87:797–806.

Ly, S., A.M. Rubenchik, S.A. Khairallah, G. Guss, and M.J. Matthews. 2017. Metal vapor micro-jet controls material redistribution in laser powder bed fusion additive manufacturing. *Nature Scientific Reports* 7(1):4085.

Mani, M., B.M. Lane, M.A. Donmez, S.C. Feng, and S.P. Moylan. 2017. A review on measurement science needs for real-time control of additive manufacturing metal powder bed fusion processes. *International Journal of Production Research* 55(5):1400–1418.

Matthews, M.J., G. Guss, S.A. Khairallah, A.M. Rubenchik, P.J. Depond, and W.E. King. 2016. Denudation of metal powder layers in laser powder bed fusion processes. *Acta Materialia* 114:33–42.

Meier, H., and C. Haberland. 2008. Experimental studies on selective laser melting of metallic parts. *Materialwissenschaft und Werkstofftechnik* 39(9):665–670.

Mercelis, P., and J.-P. Kruth. 2006. Residual stresses in selective laser sintering and selective laser melting. *Rapid Prototyping Journal* 12(5):254–265.

Mousa, A.A. 2016. Experimental investigations of curling phenomenon in selective laser sintering process. *Rapid Prototyping Journal* 22(2):405–415.

Moylan, S., E. Whitenton, B. Lane, and J. Slotwinski. 2014. Infrared thermography for laser-based powder bed fusion additive manufacturing processes. *AIP Conference Proceedings* 1581(1):1191–1196.

NIST (National Institute of Standards and Technology). 2019. Measurement Science for Additive Manufacturing Program—2018. https://www.nist.gov/programs-projects/measurement-science-additive-manufacturing-program-2018.

Niu, H.J., and I.T.H. Chang. 1999. Instability of scan tracks of selective laser sintering of high speed steel powder. *Scripta Materiala* 41(11):1229–1234.

Parry, L., I. Ashcroft, and R.D. Wildman. 2016. Understanding the effect of laser scan strategy on residual stress in selective laser melting through thermo-mechanical simulation. *Additive Manufacturing* 12(A):1–15.

Prabhakar, P., W.J. Sames, R. Dehoff, and S.S. Babu. 2015. Computational modeling of residual stress formation during the electron beam melting process for Inconel 718. *Additive Manufacturing* 7:83–91.

Puebla, K., L.E. Murr, S.M. Gaytan, E. Martinez, F. Medina, and R.B. Wicker. 2012. Effect of melt scan rate on microstructure and macrostructure for electron beam melting of Ti-6Al-4V. *Materials Sciences and Applications* 3(5):259–264.

Read, N., W. Wang, K. Essa, and M. Attallah. 2015. Selective laser melting of AlSi10Mg alloy: Process optimisation and mechanical properties development. *Materials and Design* 65:417–424.

Repossini, G., V. Laguzza, M. Grasso, and B.M. Colosimo. 2017. On the use of spatter signature for in-situ monitoring of laser powder bed fusion. *Additive Manufacturing* 16:35–48.

Scharowsky, T., V. Juechter, R.F. Singer, and C. Körner. 2015. Influence of the scanning strategy on the microstructure and mechanical properties in selective electron beam melting of Ti–6Al–4V. *Advanced Engineering Materials* 17(11).

Seyda, V., N. Kaufmann, and C. Emmelmann. 2012. Investigation of aging processes of Ti-6Al-4 V powder material in laser melting. *Physics Procedia* 39:425–431.

Spears, T.G., and S.A. Gold. 2016. In-process sensing in selective laser melting (SLM) additive manufacturing. *Integrating Materials and Manufacturing Innovation* 5(1):1–25.

Strano, G., L. Hao, R.M. Everson, and K.E. Evans. 2013. A new approach to the design and optimisation of support structures in additive manufacturing. *The International Journal of Advanced Manufacturing Technology* 66(9–12):1247–1254.

Tammas-Williams, S., H. Zhao, F. Léonard, F. Derguti, I. Todd, and P.B. Prangnell. 2015. XCT analysis of the influence of melt strategies on defect population in Ti–6Al–4V components manufactured by selective electron beam melting. *Materials Characterization* 102:47–61.

Tapia, G., and A. Elwany. 2014. A review on process monitoring and control in metal-based additive manufacturing. *Journal of Manufacturing Science and Engineering* 136(6):060801.

Thijs, L., F. Verhaeghe, T. Craeghs, J. Humbeeck, and J.-P. Kruth. 2010. A study of the microstructural evolution during selective laser melting of Ti–6Al–4V. *Acta Materialia* 58(9):3303–3312.

Thomas, D. 2009. The development of design rules for selective laser melting [Thesis]. University of Wales, Cardiff, Wales, United Kingdom.

Tolochko, N.K., S.E. Mozzharov, I.A. Yadroitsev, T. Laoui, L. Froyen, V.I. Titov, and M.B. Ignatiev. 2004. Balling processes during selective laser treatment of powders. *Rapid Prototyping Journal* 10(2):78–87.

Trushnikov, D., L.N. Krotov, E.L. Krotova, and N.A. Musikhin. 2016. Reconstructing the melting channel shape in electron-beam welding from the parameters of penetrating X-ray radiation. Part I. Posing inverse geometrical problem. *Russian Journal of Nondestructive Testing* 52(10):576–582.

Van Belle, L., G. Vansteenkiste, and J.C. Boyer. 2013. Investigation of residual stresses induced during the selective laser melting process. *Key Engineering Materials* 554–557:1828–1834.

Van Elsen, M. 2007. Complexity of selective laser melting [Ph.D. thesis]. Katholieke Universiteit Leuven.

Weingarten, C., D. Buchbinder, N. Pirch, W. Meiners, K. Wissenbach, and R. Poprawe. 2015. Formation and reduction of hydrogen porosity during selective laser melting of AlSi10Mg. *Journal of Materials Processing Technology* 221:112–120.

Yasa, E., J. Deckers, T. Craeghs, M. Badrossamay, and J.-P. Kruth. 2009. Investigation on occurrence of elevated edges in selective laser melting. Pp. 180–192 in *2009 Proceedings of the Solid Freeform Fabrication Symposium*. 20th Annual International Solid Freeform Fabrication Symposium, Austin, Tex., August 3–5.

Zaeh, M.F., and G. Branner. 2010. Investigations on residual stresses and deformations in selective laser melting. *Production Engineering* 4(1):35–45.

Zaeh, M.F., and M. Kanhert. 2009. The effect of scanning strategies on electron beam sintering. *Production Engineering* 3(3):217–224.

Zäh, M.F., and S. Lutzmann. 2010. Modelling and simulation of electron beam melting. *Production Engineering* 4(1):15–23.

Zeng, K. 2015. Optimization of support structures for selective laser melting [Ph.D. thesis]. University of Louisville, Louisville, Ky.

Zhou, X., X. Liu, D. Zhang, Z. Shen, and W. Liu. 2015. Balling phenomena in selective laser melted tungsten. *Journal of Materials Processing Technology* 222:33–42.

# 3

# Microstructure Evolution, Alloy Design, and Part Suitability

The second session of the workshop focused on developing models to represent microstructure evolution, alloy design, and part suitability. Lyle Levine (National Institute of Standards and Technology [NIST]) and Kyle Johnson (Sandia National Laboratories) gave opening presentations and were joined by Annett Seide (MTU Aero Engines), Eric Jägle (Max Planck Institute), Deniece Korzekwa (Los Alamos National Laboratory), Christian Leinenbach (Empa), and John Turner (Oak Ridge National Laboratory) for a panel discussion relating to the following questions:

- How does the additive manufacturing (AM) community develop and validate computer models that use measured material property data and build parameters to predict the location-dependent state of as-built and post-processed components?
- How does the AM community develop and validate computer models that connect the location-dependent state of a part to its performance?

## MEASUREMENTS FOR ADDITIVE MANUFACTURING OF METALS

*Lyle Levine, National Institute of Standards and Technology*

Levine began his presentation with a discussion of how the processing, structure, property, and performance stages in AM interact. He explained that feedstock material and other environmental considerations can combine with the complex build process to create a complex composition and thermal stress history. This information can inform models of residual stresses and microstructure, which can then provide estimates of mechanical properties and life-cycle behavior.

Levine provided measurements for laser powder-bed fusion, each categorized by model inputs, model guidance, and model validation (see Table 3.1). He noted that a previous AM workshop (see NASEM, 2016) stressed the importance of benchmark measurements for comparison testing. In response, NIST began the AM Benchmark Test Series (AM-Bench) and now has a scientific committee that includes 60 organizations and 83 members.[1]

Several issues arose as this committee first attempted benchmark measurements, Levine explained. First, there was a tremendous range of additive processes and materials as well as unexplained build variability between machines and processes. Time-intensive metrological-level measurements were needed, and the systems were still being built. To streamline the process, two general sets of benchmarks were made for metals. The first set of benchmarks involved 21 scientists from 6 organizations and focused on part deflection, residual elastic strains, microstructure, phase fractions, and phase evolution. The second set of benchmarks involved 14 scientists from 2 organizations and focused on low-level phenomena, including melt-pool geometry, cooling rate, topography, grain structure, dendritic microstructure, and three-dimensional structure. For a blind benchmark challenge, there were 46 submissions (almost all with metals). Levine noted that the groups that used more physics for their submissions ended up being closer to actual measurements. During a later discussion period, a participant asked Levine why there were so few valid submissions. Levine responded that the relationship between residual stress measurements and part distortion models posed challenges. There were two submissions tied for first place for predicting residual stress measurements accurately and no winner for predicting part distortions. The groups also struggled with predicting surface topography, such as chevron patterns that form on the surface of materials, and anticipating the liquid flow during the solidification process. He speculated that this could be due to surface tension issues. Microstructure evolution was also challenging, particularly in understanding what phases and precipitate sizes/shapes happen as a function of time. Few groups submitted their results for microstructure evolution.

---

[1] For more information about NIST's AM-Bench, see https://www.nist.gov/ambench, accessed October 26, 2018.

**TABLE 3.1** Measurement List for Powder-Bed Fusion

| Model inputs | Model guidance | Model validation |
|---|---|---|
| **Thermophysical parameters**<br>• Liquidus, solidus temps<br>• Latent heats<br>• Specific heat capacity<br>• Surface tension<br>• Etc.<br><br>**Build parameters**<br>• Scan pattern, power, speed<br>• Power distribution function<br>• Local cooling rates<br>• Etc.<br><br>**Part characterization**<br>• Dislocation density<br>• Phases, precipitates<br>• Etc. | **Melt pool** (in-situ builds and tracks)<br>• Length, width<br>• Absorptivity (as function of time)<br>• Cooling rate<br>• Mode (conduction or keyholing)<br>• Etc.<br><br>**Laser tracks and build layers**<br>• Widths, cross sections<br>• Grain shapes, orientations<br>• Texture<br>• Phases, precipitates<br>• Solidification microstructures<br>• Elemental segregation<br>• Etc. | **Part level** (as-built and processed)<br>• Dislocation density<br>• Phases, precipitates<br>• Microstructure evolution<br>• Texture<br>• Residual stresses/strains<br>• Part geometry<br>• Distortion<br>• Mechanical properties<br>• Fatigue properties<br>• Corrosion properties<br>• Etc. |

SOURCE: Lyle Levine, National Institute of Standards and Technology, presentation to the workshop, October 24, 2018.

Levine stated that good progress has been made on quantitative in-situ monitoring, but more development is needed for in-situ technologies. However, some needed technology is not widespread, and there is often poor traceability to primary reference standards. While the process for international benchmark measurements is under way, it is limited in scope compared to the technological need. He said that the technology for state characterization is largely developed, but some aspects are widespread and others require specialized capabilities. Lastly, he noted a severe lack of AM-compatible alloys and relevant thermophysical and related materials data.

During the question and answer portion of this presentation, Levine was asked about the current state of three-dimensional microstructure measurement. He responded that the only place he knows that does three-dimensional microstructure measurements successfully is the U.S. Naval Research Laboratory. NIST has struggled with it in the past. He explained that when doing an X-ray computerized tomography scan and looking at the grain structures, a diffraction process is being done instead of a transmission process and high dislocation densities and resolution issues appear. The only way to do three-dimensional microstructure measurement well is with a cross-sectional scanning electron microscope using localized geometry. Then, electron backscatter diffraction or other approaches can be used to examine the microstructure.

During the later panel discussion, an audience member asked about the next AM-Bench and whether Levine has thought about doing a conduction mode versus a keyhole mode. Levine stated that this was an excellent question, but the specifics need to be considered. For example, is the build conducted on one line with increasing power along that line, or is it done as a bare plate test where the power can be transitioned from conduction to keyholing? A material system also needs to be considered before transitioning.

## PREDICTING MATERIAL STATE AND PERFORMANCE OF ADDITIVELY MANUFACTURED PARTS

*Kyle Johnson, Sandia National Laboratories*

Johnson stated that AM is a multiscale, multilevel problem. Sandia National Laboratories has a vision for linking processing, structure, property, and performance via the following six programs (listed in order from short term to long term).

1. *Thermal process modeling coupled with microstructure prediction.* Sandia is working on microstructure prediction through its Stochastic Parallel PARticle Kinetic Simulator,[2] which is used for AM single continuous build and powder-bed methods.
2. *Thermal process modeling coupled with residual stress prediction.* Sandia has a Laser Engineered Net Shaping[3] process to fabricate three-dimensional metallic components directly from computer-aided design solid models and to simulate AM builds. Sandia is moving toward reduced-order models to compute the full stress states more efficiently. This process can simulate a 6-hour build time in 8 minutes. Neutron diffraction measurements are also being incorporated into performance models.
3. *Fast performance prediction accounting for as-built state, properties, and defects for qualification.* With 21 participant teams, the Third Sandia Fracture Challenge centered on predicting tensile failure of an AM part.

---

[2] For more information about Sandia National Laboratories' Stochastic Parallel PARticle Kinetic Simulator, see https://spparks.sandia.gov, accessed October 26, 2018.

[3] For more information about Sandia National Laboratories' Laser Engineered Net Shaping process, see https://www.sandia.gov/mst/technologies/net-shaping.html, accessed October 26, 2018.

4. *Efficient concurrent multiscale modeling and uncertainty quantification using techniques such as multigrid and error estimation when material statistical homogeneity does not apply.* An example would be generating microstructures using kinetic Monte Carlo, running a homogeneous simulation with an isotropic material model, recovering localized stresses using a posteriori error methods, and then comparing the results to direct numerical simulations of full kinetic Monte Carlo microstructure.
5. *Advanced high-throughput testing capability coupled with machine learning algorithms.* Full-field high-throughput testing can now be combined with machine learning. Sandia has been looking at additional volume correlation techniques to get more volumetric results instead of surface-level results. The volumetric results could then be turned into a neural network that can help determine the correlation with failure or critical defect structure.
6. *Process parameter-dependent microstructure prediction leading to local texture control and optimization.* Johnson provided an example that illustrated how process settings affect microstructure. Coupling the process-dependent microstructure and a design optimization code, such as Plato,[4] might lead to the creation of a site-specific optimized microstructure (Popovich et al., 2017). Johnson said that this could be a "game changer" but is likely still years away.

Johnson noted that challenges remain with each of these six steps. For thermal process modeling coupled with microstructure prediction, better three-dimensional microstructure imaging capabilities are needed, and representation of local microstructure on full-size parts is both a computing power and data storage issue. For thermal process modeling coupled with residual stress prediction, residual stress is still difficult to measure, type-II residual stress is difficult to predict, and an optimization for residual stress is needed. Fast performance prediction accounting for as-built states, properties, and defects for qualification still has to include uncertainty quantification for these materials. Concurrent multiscale modeling and uncertainty quantification using techniques such as multigrid and error estimation can be expensive and difficult. Crystal plasticity models need to account for as-built dislocation structures and other microstructural characteristics that are unique to AM. Lastly, Johnson noted that advanced high-throughput testing capabilities coupled with machine learning algorithms still face questions such as what to use for speckle patterns and which defects or defect networks matter, as well as how to find them.

---

[4] For more information about Sandia National Laboratories' Plato environment, see https://sierradist.sandia.gov/Plato_index.html, accessed October 26, 2018.

## DISCUSSION

Following the presentations, Jägle, Leinenbach, Korzekwa, Seide, Turner, and Johnson participated in a panel discussion on microstructure evolution, alloy design, and part suitability. In addition to the two session questions outlined at the beginning of this chapter, Levine, the moderator, posed the following questions:

- What thermophysical parameters are most needed and how can they be measured?
- AM-Bench can only provide a limited amount of data. What future benchmark measurements should have the highest priority?
- It has been suggested that transition states/instabilities are important to investigate—for example, the onset of keyholing and dimensional instabilities for thin walls. What other transition states merit investigation?
- How can commercial in-situ process monitoring systems be validated?
- What is the best role for high-performance computing in AM simulation?
- What are short-, intermediate-, and long-term needs and directions in AM?

An audience member asked the panelists to share their thoughts on microstructure evolution modeling. In particular, since the microstructure cannot be truly predicted, could a blind prediction be used as the next step? Levine responded that AM-Bench 2018 did ask what phases develop in Inconel 625 during a residual stress heat treatment. One AM-Bench group correctly predicted the phases, but the growth rate and the shape of the precipitates were incorrect. Seide stated that although blind predictions may someday be useful, they are not possible yet. Johnson added that for certain materials, such as austenitic stainless steel, predictions are fairly reasonable. However, more research is needed to understand the impact of defects. Jägle noted that blind predictions for areas such as precipitate nucleation or growth rate, where predictions are determined by defects, are currently unavailable. However, very few people go beyond classical nucleation approaches. Another question was how to better integrate sensing data into models to improve predictions. Korzekwa stated that sensing data could be used both as a model input and to validate the model output. Understanding the boundary conditions of the situation is also very important; however, this all depends on the model and what the data actually are. An audience member asked how data could be used with models to predict or estimate the full state of a system that cannot be measured directly. Levine responded that projects should generally be joint measurement and modeling efforts because one cannot model and measure everything. Measurements can be used to constrict model parameters and help identify underlying physics.

A participant asked whether a heat treatment could be devised to encourage a particular long-term microstructure evolution regardless of the as-built microstructure. Johnson responded that he is not sure if it is possible to do so. Jägle replied that there are limited options for changing the heat treatment, which is why it is important to understand the solidification process. Levine stated that developing a heat treatment for a specific AM alloy is complex. He described two cases in which NIST tried to develop heat treatments with unexpected complications. In one case, a residual stress heat treatment was needed before cutting the parts off of the build plate. This process resulted in significant amounts of unpredicted niobium, which had to be eliminated.

A participant asked about the roles of creep, fatigue, and tensile properties in microstructure evolution. Korzekwa responded that, overall, predicting segregation and texture is challenging. Temperature-dependent mechanical properties are not understood well enough to predict some of the previously mentioned heat treatment effects. She noted that more work is needed to improve modeling capabilities and estimates of relevant material properties; Jägle added that these advances could help researchers achieve desired microstructures and better understand performance. Once there is a microstructure, the models used to translate the microstructure into thermomechanical properties are similar, with additional considerations such as defects that are not present in other materials. Levine gave an example in which his team at NIST tested about six different annealing treatments before the precipitation process. The team did mechanical testing on treated parts; although these parts were composed of the same material and were subjected to heat treatments in similar ranges, tensile tests varied by a factor of three. This difference was due to microstructure variations, including which precipitates formed (and their size and predictability).

The same audience member asked how current knowledge of microstructure modeling could be applied to multicomponent alloy design. Jägle responded that there is no single approach to alloy design; it depends on the type of alloy. If he was asked to design a better aluminum alloy, he would need to design better precipitates or compositions that would work in AM.

In response to a question about model uncertainties and validation tests, Turner stated that confidence in a model is needed before exploring factors such as surface tension at various temperatures. Teter noted that sensitivity analysis of certain parameters, such as recoil versus Marangoni in the melt-pool behavior, is a big open question in AM.

Levine asked the panelists about short-, intermediate-, and long-term goals for AM of metals, which are described below.

- *Short-term goals*
    — Improving microstructure modeling, particularly for the prediction of grain size, phases, and defects (Johnson);
    — Using machine learning on in-situ monitoring data (Johnson);
    — Developing guidelines for qualification design (Johnson);
    — Modifying existing alloys to work in AM (Jägle);
    — Improving the understanding of the physics behind some materials' behaviors (Korzekwa);
    — Refining standards (Seide); and
    — Obtaining temperature-dependent thermophysical properties needed for simulations—some software systems have temperature as a function of part-geometry and other properties, which may be a direction worthy of further exploration (Leinenbach).

- *Intermediate-term goals*
    — Simulating all laser passes with computationally efficient approaches (Johnson);
    — Improving topology optimization and location-specific process optimization (Johnson);
    — Strengthening the understanding of modeling capabilities, such as process-continuous models and microstructure models (Korzekwa);
    — Expanding training in computational materials engineering (Seide); and
    — Developing a multiphysics approach for coupling capabilities (Seide).

- *Long-term goals*
    — Combining digital volume correlation with machine learning to minimize failure (Johnson);
    — Creating AM-specific alloys with specialized cooling rates (Jägle);
    — Developing more user-friendly models (Korzekwa);
    — Improving the understanding of microstructures in different parts and positions for localized needs (Seide);
    — Strengthening model reliability to predict distortion, microstructure, and mechanical properties (Leinenbach);

- Establishing a set of community models and interfaces between the different components, where lower- and higher-fidelity models can be interchanged—this could be similar to an open source version built in a collaborative environment (Turner); and
- Developing community standards on the models and interfaces (Turner).

## REFERENCES

NASEM (National Academies of Sciences, Engineering, and Medicine). 2016. *Predictive Theoretical and Computational Approaches for Additive Manufacturing: Proceedings of a Workshop.* Washington, DC: The National Academies Press.

Popovich, V., E. Borisov, A.A. Popovich, V. Sufiiarov, D.V. Masaylo, and L. Alzina. 2017. Impact of heat treatment on mechanical behaviour of Inconel 718 with tailored microstructure processed by selective laser melting. *Materials and Design* 131:12–22.

# 4

# Process and Machine Design

The third session of the workshop focused on modeling aspects of process and machine design in additive manufacturing (AM). Tahany El-Wardany (United Technologies Research Center), Ade Makinde (General Electric [GE] Global Research Center), Johannes Henrich Schleifenbaum (Fraunhofer Institute for Laser Technology), Shoufeng Yang (KU Leuven), Jian Cao (Northwestern University), Ranadip Acharya (United Technologies Research Center), Mustafa Megahed (ESI Group), and Michael Schmidt (Friedrich-Alexander Universität Erlangen-Nürnberg) each discussed research, challenges, and future directions relating to the following questions:

- How can processing and post-processing be changed to drive part and manufacturing performance to a predetermined goal (e.g., target state and production rate)?
- How can modified machine instructions bring about the desired process changes?
- What new methods or techniques need to be developed to run the AM process so that control signals can be included?
- How can part–process planning be optimized?
- What new methods or techniques for hybrid or autonomous machines need to be developed to enable real-time monitoring and control?

## MODELING PHASES OF PROCESS AND MACHINE DESIGN

*Tahany El-Wardany, United Technologies Research Center*

El-Wardany began her presentation with an overview of the United Technologies Research Center and its work in aerospace, machine design, and (most recently) AM. The company is currently focused on researching and optimizing AM processes and exploring different modeling techniques.

El-Wardany discussed AM part processing, modeling and validation of AM processes, current AM machines, and requirements for AM machine design and capabilities. She noted that there are limitations and gaps for existing machine design; challenges in controlling machine performance, reproducibility, and repeatability; needs for closed-loop monitoring and the ability to output controller data; needs for multifidelity modeling of processes to influence selection of monitoring strategy and signature; and needs for full part modeling and optimization of scan strategy.

When making an AM part, she explained, there are three distinct phases: pre-processing, process selection, and post-processing. During pre-processing, the part concept is translated into a computer-aided design model. This model is then used to improve and, hopefully, to optimize the AM processes. Pre-processing also involves the preparation of parameters such as support generation, orientation, build layout, nesting, and scan strategy. After these parameters are defined, it is time to move to the process selection phase, which includes processing, monitoring, and control. After this phase, there should be enough information for post-processing practices such as unpacking, part cleaning, stress relief, part removal, support removal, heat treatment and hipping, surface finishing, and part inspection.

It is important to discuss the limitations and gaps within these three phases, El-Wardany explained. New models are needed to advance the design of an AM part, specifically for generative design, cost projection, parametric modeling, and multiphysics optimization. She also mentioned the increased availability of pre-processing software for parameter selection, scan strategy, and build-file generation (e.g., Magics Build Processor and Machine). With current modeling and validation of AM processes, the objectives are to (1) develop integrated physics-based simulation tools of AM processes to predict part-level distortion, defects, and microstructure as well as to establish correlation to performance (i.e., fatigue); and (2) use the developed tools to reduce AM process development time and cost. These processes include powder-bed fusion, powder-directed energy deposition, and wire arc AM. El-Wardany discussed important considerations for validating high-fidelity physics-based models used to predict properties and part life, such as heat source, melt-pool dynamics, thermal history, bulk distortion, and microstructure evolution. Power and speed are the main variables that can help map processes. There is still a need for more modeling activity to optimize heat treatment and predict part life. El-Wardany presented a chart on available AM processes and equipment (shown in Table 4.1), provided a few examples of current machines,[1] and noted that new machines and applications are progressing every year. She also discussed the following limitations for these processes:

- *Fused deposition modeling*: Weak mechanical properties, limited materials (only thermoplastics), and inconsistent surface finish.
- *Powder-bed fusion (selective laser sintering, selective laser melting, and electron beam melting)*: Slow printing and high cost.
- *Inkjet printing and contour crafting*: Difficulty maintaining workability, coarse resolution, lack of adhesion between layers, and inconsistent surface finish.
- *Stereolithography*: Very limited materials, slow printing, and high cost.
- *Directed energy deposition*: Poor accuracy, low surface quality, need for a dense support structure, and limitation in printing complex shapes with fine details.
- *Laminated object manufacturing*: Poor surface quality and dimensional accuracy and limitation in the manufacturing of complex shapes.
- *Part size*: Lack of novel approaches to relieve stresses and distortion for large-scale parts.
- *Scalability*: High machine and material costs.
- *Limited material and high cost*: Lack of affordable AM-adapted materials.
- *Inconsistent quality*: Part quality is difficult to control, machine-to-machine repeatability and reproducibility of parts are a challenge, accessing machine controller for feedback and process modifications is limited, and in-situ sensing and monitoring systems are rarely available.
- *AM machines*: Lack of examples in multifunctional structures, functionally graded materials, and automated repair processes.

---

[1] Example machines include the Selective Laser Melting Machine 280, Matsuura LUMEX, DMG Mori LASERTEC 65, Friction surface AM Aeroprobe, BeAM Modulo 400, FDM Fortus 450mc, and Polyjet Stratasys j750.

**TABLE 4.1** Available AM Processes and Equipment

| Process category | Process or technology | Material | Manufacturer | Machine |
|---|---|---|---|---|
| Vat photo-polymerization | Stereolithography | Ultraviolet curable resins | Asiga | Freeform Pico |
| | | | 3D Systems | iPro, Projet6000/7000 |
| | | | EnvisionTEC | Perfactory |
| | | | Rapidshape | S Series |
| | | Waxes | DWS | DigitalWax |
| | | Ceramics | Lithoz | CeraFab 7500 |
| Material jetting | Multijet modeling | Ultraviolet curable resins | 3D Systems | Projet 3500 HD/3510/ 5000/5500 |
| | | | Stratasys | Objet |
| | | Waxes | Solidscape | 3Z |
| Binder jetting | 3D printing | Composites | 3D Systems | Z Printer |
| | | Polymers, ceramics | Voxeljet | VX Series |
| | | Metals | ExOne | M-Flex |
| Material extrusion | Fused deposition modeling | Thermoplastics | Stratasys | Dimension, Fortus, Mojo uPrint |
| | | | MakerBot | Replicator |
| | | | RepRap | RepRap |
| | | | Bits from Bytes | 3D Touch |
| | | | Fabbster | Fabbster Kit |
| | | | Delta Micro Factory Corp. | UP |
| | | | Beijing Tiertime | Inspire A450 |
| | | Waxes | Choc Edge | Choc Creator V1 |
| | | | Essential Dynamics | Imagine |
| | | | Fab@Home | Model |
| | | Metal | nScrypt | 3DnþnMill Three-axis CNC machine |
| | | | Hyrel 3D | Hydra 340, 640, 645 3-axis CNC machining and laser cutting |

*continued*

**TABLE 4.1 Continued**

| Process category | Process or technology | Material | Manufacturer | Machine |
|---|---|---|---|---|
| Powder-bed fusion | Selective laser sintering | Thermoplastics | EOS | EOS P |
| | | | Blueprinter | Selective heat sintering |
| | | | 3D Systems | sPro |
| | | Metals | 3Geometry | DSM |
| | | | Matsuura | Lumex Avance-25 and 60 3-axis CNC machining controlled atmosphere |
| | Selective laser melting | Metals | 3D Systems/Phenix | PXL, PXM, PXS |
| | | | EOS | EOSINT M |
| | | | SLM Solutions | SLM |
| | | | Concept Laser | LaserCusing |
| | | | 3D Systems | ProX |
| | Electron beam melting | Metals | Realizer | SLM |
| | | | Renishaw | AM250 |
| | | | Arcam | Arcam A2 |
| | | | Sciaky | DM |
| Sheet lamination | Laminated object manufacturing | Paper polymers | Mcor | Technologies Matrix 300þ |
| | | Metals | Fabrisonic | SonicLayer |
| | | Thermoplastics | Solido | SD300Pro |

**TABLE 4.1 Continued**

| Process category | Process or technology | Material | Manufacturer | Machine |
|---|---|---|---|---|
| Directed energy deposition | Laser metal deposition or laser engineered net shaping | Metal | Optomec<br>DM3D<br>Irepa Laser | LENS 450, LENS 3D (hybrid system—5 axis CNC machine-controlled atmosphere) |
| | Electron beam AM | Metal | Sciaky | Directed Metal Deposition, EasyCLAD, VX-110 |
| | | | | Robotic-based applications |
| | Wire arc AM | Metal | DMG MORI | LASERTEC 65 3D, LASERTEC powder nozzle and powder bed |

NOTE: 3D, three dimensional; CNC, computer numerical control; DM, digital metal; SLM, selective laser melting.
SOURCE: Tahany El-Wardany, United Technologies Research Center, presentation to the workshop, October 25, 2018.

## CURRENT STATE OF COMMERCIAL POWDER-BED ADDITIVE MACHINES—AM MACHINE DESIGN ISSUES IMPACTING BUILD-TO-BUILD AND PART-TO-PART VARIABILITY

*Ade Makinde, General Electric Global Research Center, with support from Johannes Henrich Schleifenbaum, Fraunhofer Institute for Laser Technology, and Shoufeng Yang, KU Leuven*

Makinde described GE Additive, which was launched in 2016 and includes divisions such as AddWorks™ consultancy, machine modalities, advanced powders and coating materials, software, and customer experience centers. Efforts are under way to develop the world's largest additive machine with its Additive Technology Large Area System (Project A.T.L.A.S.). He noted that GE is examining how to use AM across industries and is committed to having 25 percent of its portfolio touched by AM by 2025.

Within GE, Makinde focuses on multiphysics modeling of AM processes and different tools that can be used to understand part-build quality. This involves understanding process parameters and their impacts as well as where multiphysics particle models, laser-scanning models, lattice Boltzmann methods for laser-powder interactions, and part-level models could play important roles. However, challenges include analyzing across different length scales (such as going from microns to meters), validating models, and getting codes to work together. By combining physics-based models and data-driven models, uncertainty can be quantified for part performance.

Makinde's presentation focused on AM machine design characteristics (e.g., powder-bed delivery system, laser system, chamber design, and in-situ sensors for monitoring and control) that directly impact part-build quality. The main objectives of design are to increase production rate, decrease cost, reduce defects, and meet quality requirements.

For powder-bed delivery systems, storage and environmental control of the powder are critical. He noted that moisture is an important factor that needs to be controlled when producing the powder-bed delivery system. The second aspect is the delivery of the powder and avoidance of flow separation. Makinde described the following as "low hanging fruit":

- Moving the powder throughout the machine without breaking up the powder material;
- Filtering powder without clogging;
- Finding the best technique to spread the powder;
- Examining the re-coating to see how it affects the wear, contamination, etc.; and
- Re-using powder (e.g., breaking up and mixing clumped powder).

Makinde explained that most laser systems operate on black-box control systems, but there is a need for open-source control (e.g., a G-code type). He highlighted the following important characteristics:

- Power;
- Speed;
- Response time;
- The time at which a machine is coded or reprogrammed (the build can sometimes take 4 to 6 weeks);
- Galvo, the laser power and scanning mirror speed control;
- Inclination of the scan angles (as build chamber size increases);
- Thermal lensing, especially for long duration builds;
- Fumes, which can be detrimental to the laser systems;
- Laser wavelength, which typically needs to be suitable for different materials; and
- Hatch pattern.

He added that most machines are using a continuous wave laser, but it is vital to understand the impact between a pulsed wave laser and a continuous wave laser.

Chamber design and oxygen levels are important characteristics for AM machines, he explained. The environmental conditions inside the chamber could affect the quality of the part. Powder-bed uniformity (i.e., particle distribution), gas flow, and soot prevention are also important. Preheating at both the build-plate and layer level can help minimize defects. Lastly, he explained that a modular chamber design for productivity, cool down, and de-powdering can help increase productivity rates. The industrial control capabilities depend on sensors (e.g., pyrometer, charge-coupled device camera, height scanner), measurements (e.g., melt-pool temperature, melt-pool size, build height), stable characteristics (e.g., build height, melt-pool temperature, solidification rate, cooling rate), and outputs (e.g., laser power, powder mass, machine feed, active cooling).

For in-situ sensors for monitoring and control, Makinde said that there needs to be an integrated process and controlled environment for sensing, monitoring, and controlling thermal behavior as well as for using optical sensing for layer spread, powder spreading, melt-pool monitoring, and infrared detection.

## MODELING CHALLENGES AND OPPORTUNITIES AT THE PART LEVEL

*Jian Cao, Northwestern University, with support from Ranadip Acharya, United Technologies Research Center, and Mustafa Megahed, ESI Group*

Cao began by explaining the needs for simulations on process planning (e.g., choosing the best strategy) and material property prediction. Thermal simulation of the full component is needed to enable iteration and optimization of process planning. Thermal simulations can inform understanding of the microstructure and part distortion (which can impact the selection of laser parameters and hatch spacing), help identify hot and cool spots, and enable material property predictions of grain sizes, material phases, porosity, mechanical properties, and residual stress.

Cao showed a diagram, Figure 4.1, representing the relationships among processing, structure, properties, and performance (PSPP) in AM. Cao explained that PSPP focuses on the point level, so other considerations are needed to get to the part level. Some of these considerations include the design (e.g., product, material, and process), the machine (e.g., use of sensors and other intelligence), and the final qualification and

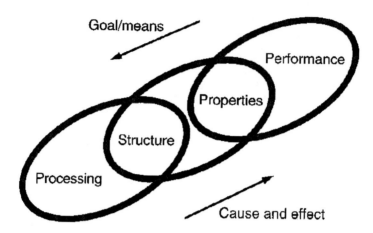

**FIGURE 4.1** Three-link chain model of the central paradigm of materials science and engineering. SOURCE: From G.B. Olson, 1997, Computational design of hierarchically structured materials, *Science* 277(5330):1237–1242. Reprinted with permission from the American Association for the Advancement of Science.

certification. The ultimate goal of an autonomous process starting from design to product would be to combine these steps.

Cao showed the critical length scales for AM products and their correspondingly normalized values for part length scales (see Table 4.2) and the critical time scales in building and using AM parts normalized with the build time scales (see Table 4.3).

There are some challenges for simulations at the part level, including the following:

- Speed and predictability, lack of failure criteria, and issues in microstructure and residual stress prediction;
- Database integration, including the extraction of useful information to be integrated into various software packages;
- Integration of powder-level and melt-pool scale models, often due to a mismatch of scales;
- Integration with pre-processing (e.g., powder spread) and post-processing (e.g., heat treatment) since the integration involves multiple processes that often have different simulation packages or different physics;
- Variability in uncertainty quantification;
- Models for in-situ process control; and
- Model validation (e.g., temperature, history, residual stress).

## PROCESS AND MACHINE DESIGN

**TABLE 4.2** Scales for Different Simulations at Critical Length Scale and Normalized Length by Part Scale

|  | Critical length scale (m) | Normalized length by that at the part scale |
|---|---|---|
| Part | $10^{-2} - 1$ | 1 |
| Feature size | $10^{-5} - 10^{-3}$ | $\sim 10^{-3}$ |
| Powder | $10^{-5} - 10^{-4}$ | $10^{-2} \sim 10^{-5}$ |
| Doping | $10^{-9}$ | $\sim 10^{-8}$ |
| Beam spot | $10^{-6} - 10^{-3}$ | $\sim 10^{-3}$ |
| Melt-pool length | $10^{-4} - 10^{-2}$ | $10^{-4} - 10^{-2}$ |
| Melt-pool depth | $10^{-5} - 10^{-3}$ | $\sim 10^{-3}$ |
| Mushy zone | $10^{-6} - 10^{-4}$ | $\sim 10^{-4}$ |
| Grain | $10^{-6} - 10^{-2}$ | $10^{-6} - 10^{-2}$ |
| Dendrite | $10^{-7} - 10^{-6}$ | $\sim 10^{-5}$ |
| Crack | $10^{-6} - 10^{-2}$ | $10^{-6} - 10^{-2}$ |

NOTE: Blue shading indicates desired to be simulated; green shading indicates needed at part level; orange shading indicates currently not simulated at the part level.
SOURCE: Gregory J. Wagner and Jian Cao of Northwestern University.


**TABLE 4.3** Scales for Different Simulations at Normalized Time by That at the Build Scale

|  | Critical time scale (sec) | Normalized time by that at the build scale |
|---|---|---|
| Part life | $10^7 - 10^9$ | $\sim 10^4 - 10^7$ |
| Build time | $10^2 - 10^4$ | 1 |
| Layer time | $10^0 - 10^2$ | $\sim 10^2 - 10^3$ |
| Solidification time scale | $10^{-4} - 10^{-3}$ | $\sim 10^{-6}$ |
| Thermal diffusion time scale | $10^{-5} - 10^{-3}$ | $\sim 10^{-7}$ |
| Thermal convection time scale | $10^{-5} - 10^{-4}$ | $\sim 10^{-7}$ |

NOTE: Blue shading indicates desired to be simulated; green shading indicates needed at part level; orange shading indicates currently not simulated at the part level.
SOURCE: Gregory J. Wagner and Jian Cao of Northwestern University.

Cao explained that there are several opportunities to address these challenges. AM-specific finite element method companies, software developers, and start-ups are working together to help design and construct parts to improve speed and predictability. With this, there is also a need for efficient surface representation. The use of graphics processing units or parallelization could lead to time improvements of several orders of magnitude (Mozaffar et al., 2019). In terms of the challenges for database and post-processing integration, there are universal file formats such as voxel representations (e.g., the Visualization Toolkit) that can integrate different models for materials thermodynamics and diffusion kinetics models for phenomenological methods for solid-state phase transformation (e.g., Computer Coupling of Phase Diagrams and Thermochemistry with the Johnson–Mehl–Avrami–Kolmogorov equation). This information can potentially be linked with other processes as well (i.e., machine operations). Surrogate models can also help with speed and predictability, database integration, and integration with powder-level and melt-pool scale models. Process maps can help with speed and predictability as well as database integration by using the absorbed power and velocity for solidification microstructure (Beuth et al., 2013). Gaussian process metamodeling and machine learning approaches can improve speed and predictability, database integration, and integration of powder-level and melt-pool scale models. Machine learning can also help to predict the thermal history using normalized temperature and time (Mozaffar et al., 2018).

After the presentation, an audience member asked Cao for her priority challenges. She stated that speed and predictability are the top priorities. In response to another audience member's question, Cao stated that understanding physics is fundamental for these simulations and helps improve predictability. Multiscale simulation tools can be developed to fully integrate or pass the critical and equivalent data from the fine scale model that incorporates detailed physics to the course scale. An audience member wondered how the graphics processing units available in many current machines would affect the types of simulations that could be done. Cao responded that these simulations could help prevent problems. Currently, in most cases, it might not be possible to fix local defects after they happen (e.g., fixing a porosity left in the previous layer); however, with more research and more data, it is possible to adjust process parameters to fix these local defects within specified limits. More importantly, it is now possible to correct some global defects, such as distortion.

## DISCUSSION

Following the presentations, Acharya, Cao, Makinde, Megahed, Schleifenbaum, Schmidt, and Yang participated in a panel discussion moderated by El-Wardany. An audience member mentioned problems with the keyhole phenomenon and asked whether it is possible to change the scale of the systems. Yang responded that it is possible to change the scale of the systems in AM but not in the design itself. And in some cases, the design needs to be modified in order to have control of the system. El-Wardany asked each panelist to comment on how to best select machines to create a desired product. Yang mentioned that concerns remain, such as how the AM process ultimately impacts the part quality and how to fix process parameters. Better monitoring and choosing the correct signals and sensors for these systems could help elucidate the relationship between the process and part quality.

Makinde asked Schmidt if there are experimental studies to guide designers on which processes to consider when making these machines. Schmidt responded that companies are working on experimental studies that change the intensity profile of certain parameters to account for factors such as energy saving and melt-pool flow, but these factors also depend on the material properties (e.g., viscosity) and temperature changes. Therefore, the intensity profiles for parameters vary in different materials. Schleifenbaum added that different approaches to preheating are beneficial. Cao mentioned that blue laser research being conducted by some U.S. companies has the potential to increase the processing speed in AM. Some cases have shown that blue lasers can increase productivity tenfold in welding. Yang stated that green laser technology still needs to be improved to enable selective laser melting. He explained that green lasers still have poor beam quality compared to 1064 nm fiber lasers, and, although the absorption rate is better for some materials like pure copper and silver, the poor beam quality gives a large focus point. Acharya stated that in order to avoid any defects, reduced-order modeling and process mapping are needed. In-line monitoring and feedback control can help better address the process map, and reduced-order models can include distortion compensation to obtain accurate geometry.

## REFERENCES

Beuth, J., J. Fox, J. Gockel, C. Montgomery, R. Yang, H. Qiao, E. Soylemez, P. Reeseewatt, A. Anvari, S. Narra, and N. Klingbeil. 2013. "Process Mapping for Qualification Across Multiple Direct Metal Additive Manufacturing Processes." Presented at the Solid Freeform Fabrication Symposium, Austin, Tex., August 12–14.

Mozaffar, M., A. Paul, R. Al-Bahrani, S. Wolff, A.N. Choudhary, A. Agrawal, K. Ehmann, and J. Cao. 2018. Data-driven prediction of the high-dimensional thermal history in directed energy deposition processes via recurrent neural networks. *Manufacturing Letters* 18:35–39.

Mozaffar, M., E. Ndip-Agbor, S. Lin, G. Wagner, K. Ehmann, and J. Cao. 2019. Acceleration strategies for explicit finite element analysis of metal powder-based additive manufacturing processes using graphical processing units. *Computational Mechanics.* https://doi.org/10.1007/s00466-019-01685-4.

# 5

# Product and Process Qualification and Certification

The fourth session of the workshop included presentations on accelerating product and process qualification and certification in additive manufacturing (AM). Paolo Gennaro (GF Precicast Additive SA), Adhish Majmudar on behalf of Michel Delanaye (GeonX), Vincent Paquit (Oak Ridge National Laboratory), Jens Telgkamp (Airbus Operations GmbH), David Teter (Los Alamos National Laboratory), and Richard Ricker (National Institute of Standards and Technology [NIST]) each discussed research, challenges, and future directions relating to the following questions:

- How can each part be built to be identical and conformant, within standard tolerances and without individual inspections?
- What new standards, methods, or techniques need to be developed to certify a part built with AM?

## PROCESS QUALIFICATION AND TECHNOLOGICAL VALIDATION, FROM CASTING TO ADDITIVE

*Paolo Gennaro, GF Precicast Additive SA*

Gennaro introduced GF Precicast Additive SA, including its three large divisions: GF Piping Systems, GF Casting Solutions, and GF Machining Solutions. GF Precicast Additive SA was founded in November 2016 and focuses on electron beam melting AM methods for titanium aluminide and titanium Ti-6Al-4V; direct metal laser sintering for nickel, cobalt, or other superalloys; and cladding for industrial materials (which is still in development). GF Precicast Additive SA has a fully certified supply chain, including the AM build, heat treatment and hot isostatic pressing, and finishing quality inspection. Gennaro discussed important steps for system qualification, process qualification, and part validation, as highlighted in Table 5.1.

One of the main advantages for improved qualification and validation would be the reduction of cost, he explained. He speculated that powders might cost less if there were only one stock for one material. An audience member noted that having a single stock might be a good short-term goal, but specific applications might need a larger suite of materials in the future. Another participant added that it depends on the company as well, since each company has a different philosophy on how it designs pieces. Gennaro responded by saying that since the final products will be similar, the standards should also be somewhat similar.

The Asset Management Standards from the International Organization for Standardization (AMS-ISO) could help establish customer standards

**TABLE 5.1** Key Milestones, Standards, and Advantages for System Qualification, Process Qualification, and Part Validation

| Task | Milestones | Customer standards referring to AMS-ISO | Advantages |
|---|---|---|---|
| System qualification | • Materials (powders)<br>• Equipment (electron beam melting, direct metal laser melting, laser metal deposition) calibration<br>• Personnel training | • Related to AMS-ISO<br>• ISO 9100-9001 + machine training | • One stock for one material (lower cost on powders)<br>• A single machine qualification is valid for all customers |
| Process qualification | • Machine<br>• Materials<br>• Process parameters | Related to AMS-ISO | A single machine qualification is valid for all customers |
| Part validation | • Geometry on components and specimens | Acceptance criteria (X-ray, fluorescent penetrant inspection, microstructure) referring to AMS-ISO | No discussion on quality escapes |

NOTE: AMS, Asset Management Standards; ISO, International Organization for Standardization.
SOURCE: Paolo Gennaro, GF Precicast Additive SA, presentation to the workshop, October 25, 2018.

for machine qualifications. Although the milestones described in Table 5.1 are difficult to implement, doing so could help deliver a cost-effective AM process qualification and technological validation. In response to a question on in-situ monitoring, Gennaro emphasized that the goal is to complete products correctly the first time. In-situ monitoring reduces the time to fix problems that can affect the quality of a product. Makinde also mentioned that modeling and sensing could help with calibration.

## MODELING AND SIMULATION

*Adhish Majmudar, GeonX (presenting on behalf of Michel Delanaye)*

Majmudar began by referring back to Ade Makinde's presentation describing GE Additive. All of the departments focus on a vision of creating a part correctly the first time. He gave an example of a part that had problems with manufacturing, including collision, shrinkage lines, and surface defects such as small cracks. Time is lost if a part fails, and it ended up taking 24 hours to make the part from the powder. Simulations of powder-bed fusion AM are needed to address these problems, but modeling challenges remain.

Majmudar discussed how GE goes from micron-scale to part-level simulations to help design a part. GE provides a workflow to its clients where they start from a particle-bed or single-track simulation in order to look at the melt pool. Then, that information is fed into a model at a track level, which feeds into the macro-level simulation in order to predict any distortions and residual stress.

Majmudar showed a demonstration of NIST's AM Bench Challenge. Different process parameters were changed in three cases, resulting in different shapes of the melt pool. He also showed a demonstration of a mesoscale model of a track-level simulation. Nonlinear thermomechanical feeds can be determined by inputting data on the laser power, laser efficiency, laser speed, stripe angle, stripe angle increment, hatch distance, and powder material. This model matched well with experimental temperatures. Majmudar explained that thermal simulations help identify these defects or potential problems in parts. He also showed solidification models where thermomechanical properties can be used to estimate scan-level outputs, which helps to predict dendrite shapes and segregation.

Majmudar emphasized that more material properties are needed for modeling, but these can be expensive and time-consuming to obtain. Another challenge is failure prediction, specifically estimating cracking during a build. Failure during the development of a part results in significant delays in the development process, and it is difficult to understand whether a new alloy is buildable or will crack. There are also ongoing challenges related to predicting the microstructures of AM parts.

An audience member asked how to couple length scales from microscale to mesoscale. Majmudar responded that the mesoscale simulations track the laser as it moves through the melt pool. This can also be characterized by using a microscale model. Another participant asked if Majmudar performed any validation of the microstructure model. He confirmed that his team regularly does quantitative validations for thermal and mechanical distortions of the parts and discussed how a spatial grain structure model compared to experimental results. Another audience member commented on the distortion model and the quantitative validation: if the support structure were too complex to model, finding trends in the model data could help avoid the problem. In response to a question about how long simulations for the industrial part took, Majmudar explained that the turnaround time was about 24 hours, including 3 to 4 hours for the melt-pool simulation, less than 1 hour for the thermal simulation, and several hours for the full mechanical simulation.

## DISCUSSION

Following the presentations, Paquit, Telgkamp, Majumdar, Teter, and Ricker participated in a panel discussion led by Gennaro. An audience member noted that the other side of verification is making sure the machine is reliable and cannot be corrupted. She asked the panelists whether any of them are also considering approaches toward improving assurance, trust, and security. Teter mentioned that Los Alamos National Laboratory is considering these approaches in its work. While assurance typically means that all of the parts meet the requirements to be certified, not enough may be known about the process, structure, and property requirements to reach this goal. Gennaro discussed doing a risk assessment to help with mass production of parts. Ricker mentioned that there was a workshop (see Williams, 2015) at NIST about cybersecurity for print digital manufacturing in which a speaker had students build sample parts while he hacked their codes and put in defects without their knowledge. Ricker stated that machine hacking is a vulnerability for AM, as it is in all types of cyber-physical systems. He suggested that one could use a separate monitoring system that is independent from the computer and facilities that are doing the build to prevent both systems from being affected by the same hack.

John Turner (Oak Ridge National Laboratory) noted that conventional manufacturing has vulnerabilities as well, particularly when only one domestic supplier exists. He asked the panelists whether there are opportunities for AM to increase overall trust in the supply chain. Telgkamp stated that a long-term vision is to go from a classical supply chain with specialized suppliers to a system where one of many possible suppliers could be identified to produce a part using AM. Makinde mentioned that GE's software includes a blockchain feature to ensure the stability of a frozen process. Paquit commented that blockchain is not going to address cybersecurity challenges inside the machine, but sensors may be able to help with that. Telgkamp said that blockchain could be helpful in the future to attach the digital proof of quality to an individual part.

An audience member mentioned the lack of standards for safety of parameters, especially for powders, and asked whether qualification standards exist. Telgkamp said that his group has safety documents in place for mandatory requirements from suppliers. Gennaro stated that the suppliers need to provide a safety data sheet since they best understand the powder and how to use it safely. Teter mentioned that standards for testing the flammability of powders exist, but it can be difficult to find the facilities and resources to perform the tests. Another audience member asked about the possibility of reusing powder. Telgkamp replied that there needs to be a verified process and systematic investigation in place. Ricker and Teter added that water vapor, nitrogen levels, and corrosion are important considerations for powder reuse.

A participant asked whether standards exist to address defects that are rare but catastrophic. Paquit answered that sensing may be a short-term solution to avoid issues that result from defects. Teter mentioned that he thinks about the critical flaw size and location of common defects since some areas are more sensitive to defects than other areas within the part.

Teter asked about the use of model validation for instances when researchers can predict a result such as a mechanical property but cannot change any parameters. Majmudar stated that this question leads to discussions of variability in the process, which could also help researchers better understand measurement errors and common causes for variability. Ricker added that many tests are currently required to assess variability for qualification, and models can help understand variability and build trust in the systems. Teter emphasized the importance of representing the underlying physics of the materials and mechanical properties in machine learning models to increase the meaningfulness of possible predictions.

The panelists elaborated on the use of experimental data and modeling for calibration and qualification in response to a question from the audience. Gennaro stated that his team uses experimental data to help with calibrations. Teter emphasized that modeling is helpful in the qualification process, particularly with understanding which parameters are most sensitive to part quality. Modeling can help guide and focus the experimental efforts.

Gennaro asked how long it would take to certify the in-situ monitoring approaches to generate one part of production if computerized tomography (CT) scans were involved. An audience member noted that CT measurements are not used on their own because they are not traceable, unlike other measurements used in certification.

An audience member noted that more data are not always better for certification. Ricker agreed, particularly for data collected early in the process that may not be as relevant to the final part. Teter stated that data from a part with a known defect could be compared to data for other parts to help understand the impact of the defect, but this comparison depends on how well the sensors measure important parameters. Paquit added that his team stores a large amount of data to help address future questions and that it is important to have diversity in the data. Another participant commented that it is important to learn how to use these data to support decision making; in the future, hopefully all data will be usable for production.

## REFERENCE

Williams, C.B. 2015. "An Analysis of Cyber Physical Vulnerabilities in Additive Manufacturing," Pp. 21–50 in *Proceedings of the Cybersecurity for Direct Digital Manufacturing (DDM) Symposium* (C. Paulsen, ed.). NISTIR 8041, National Institute of Standards and Technology, Gaithersburg, Md.

# 6

# Summary of Challenges from Subgroup Discussions and Participant Comments

During the third day of the workshop, participants met in subgroups to discuss some of the challenges in additive manufacturing (AM). These groups aligned with the four sessions of the workshop:

1. Measurements and modeling for process monitoring and control;
2. Developing models to represent microstructure evolution, alloy design, and part suitability;
3. Modeling aspects of process and machine design; and
4. Accelerating product and process qualification and certification.

Breakout groups were asked to discuss two or three principal topics and consider the following overarching questions:

- What are the greatest technological challenges?
- What are the most important areas for research?
- What are "nontechnical" challenges to commercialization of AM?
- How can industry and academia better interact and collaborate to address technical and nontechnical challenges?
- Are there concrete actions that could help address the challenges identified?
- What topics could be addressed in a follow-on workshop?

Workshop participants were also asked to provide individual responses to similar questions about top priority research needs for advancing AM, top "nontechnical" challenges to commercialization of AM, and actions that could help address these nontechnical challenges. Descriptions of the subgroup discussions and individual responses are provided in the following subsections.

## MEASUREMENTS AND MODELING FOR PROCESS MONITORING AND CONTROL

*Subgroup Members*

Jarred Heigel (National Institute of Standards and Technology), Carolin Körner (Friedrich-Alexander Universität Erlangen-Nürnberg), Amit Surana (United Technologies Research Center), R. Allen Roach (Sandia National Laboratories), Kilian Wasmer (Empa), Shoufeng Yang (KU Leuven), and Celia Merzbacher (SRI International)

### Breakout Discussion

This breakout group discussion was led by Heigel, and conversations focused on sensor technology, algorithm development and use, knowledge transfer, challenges, and priorities moving forward. The following three questions were proposed to start the discussion:

1. *What is good enough? How much information is needed from the process to meet the desired goals?* Some subgroup members noted that clearly identifying what process information is needed will enable the development of useful sensors.
2. *What information must be exchanged between real-time monitoring sensors and process models?* Several subgroup members commented that this is specific for model-based control and is an immediate need.
3. *How can decisions and guidelines be made for processing and saving measurements?* Many subgroup members commented that this question also addresses issues of data management.

Several subgroup members highlighted the following technical challenges:

- *Correlating process phenomena with structures and defects and incorporating real data into process models.* This could help improve the understanding of the overall AM process and the underlying physics (e.g., understanding what may increase the chance of failures or unsatisfactory parts), which in turn could help improve the sensor design and the data analysis.

- *Clarifying what needs to be measured to control the outcome.* This may include defining the industrial needs for real-time monitoring.
- *Understanding the material, structure, and defect specifications.* This is particularly important in terms of understanding areas of concern within the part regarding defects and what defect density is acceptable. It is important to be able to define what is and is not acceptable for specific parts and design criteria.
- *Providing better input and output definitions for models and sensors.* This could help to improve communication throughout the system. While each sensor represents a different aspect of the process, they can provide a more complete picture of the process when they are combined.
- *Assessing whether sensor systems are capable of measuring critical parameters and providing real-time analysis.* It is important to question what hardware and analysis are needed if current systems are not fast enough to enable sufficient process control.
- *Enabling the long-term goal of a feed-forward loop based on reliable models.* This is a significant challenge that is also dependent on the previously mentioned challenges.

Subgroup members discussed the challenge of machine interoperability and how to encourage machine manufacturers to be more transparent with their systems and processes. Currently, the high cost of developing these systems and the associated intellectual property deters manufacturers from making their systems more transparent. However, many manufacturers are small organizations that may lack the resources and expertise required to develop real-time monitoring and process control strategies required by the end user. On the other end of the spectrum, organizations with monitoring and control expertise often are not as familiar with the intricacies of the process and lack the ability to communicate directly with the machines. Some members of the subgroup speculated that case studies and cost analysis could help to convince manufacturers that increasing the transparency of their machines and enhancing collaboration will serve the greater good of the AM community and ultimately increase manufacturers' customer base. The semiconductor industry—which has benefited from collaborations and partnerships—could be an exemplar of how to encourage transparency and collaboration among small companies. Finally, many subgroup members suggested that a follow-on workshop could focus on data collection and improved decision making.

# DEVELOPING MODELS TO REPRESENT MICROSTRUCTURE EVOLUTION, ALLOY DESIGN, AND PART SUITABILITY

*Subgroup Members*

*Annett Seide (MTU Aero Engines), Lyle Levine (National Institute of Standards and Technology), John Turner (Oak Ridge National Laboratory), Ade Makinde (General Electric Global Research Center), Kyle Johnson (Sandia National Laboratories), Eric Jägle (Max Planck Institute), Deniece Korzekwa (Los Alamos National Laboratory), and Christian Leinenbach (Empa)*

## Breakout Discussion

This breakout group discussion was led by Seide, who noted that many of the challenges in representing microstructure evolution, alloy design, and part suitability are encompassed by the larger material science research effort. However, there are unique areas of ongoing research that are specific to AM materials and conditions. The subgroup first discussed the lack of thermophysical data under AM conditions, and several members suggested the following short-, intermediate-, and long-term goals:

- *Short-term goals*: Identify the data that are needed for process measurement and for modeling input.
- *Intermediate-term goals*: Obtain data for a limited set of AM materials.
- *Long-term goals*: Take a deep look at the quality of data and explore first principles and machine learning approaches.

These members emphasized that the most important areas of research for the lack of thermophysical data are first principles and machine learning approaches.

The subgroup next discussed microstructure evolution and the challenge of developing and validating models. In particular, several members described high-fidelity models, coupled multiphysics models (e.g., to get location-specific microstructure evolution and to look through the solidification and intrinsic heat treatment processes as well as post-build processing), and reduced-order models as particularly challenging. Many subgroup members noted several promising short-, intermediate-, and long-term research areas:

- *Short-term research areas*: Conducting sensitivity analysis of current model parameters, coupling physical phenomena in models,

temporal and spatial scale bridging, and three-dimensional microstructure characterization information.
- *Intermediate-term research areas*: Modeling phenomena of interest under nonequilibrium conditions and calibration of model parameters.
- *Long-term research areas*: Predicting metastable phases, predicting models for nucleation, and predicting interfacial energies.

The third topic discussed was coupled multiphysics and multiscale capabilities for AM, including the laser-material interaction, the time-dependent thermal profile (including fluid flow), microstructure evolution, micromechanics, and macroscale thermomechanics. Many subgroup members noted several promising short-, intermediate-, and long-term research areas:

- *Short-term research areas*
  — Analysis of coupling between relevant physics. This is challenging because it requires a fully coupled model.
  — Exploration of approaches for modeling laser-scan strategies. These approaches may be done through parallel in-time approaches.
  — Prediction of site-specific properties. This includes properties throughout the parts and the ability to use site-specific properties in macroscale models.
- *Intermediate-term research areas*
  — Development of reduced-order models informed by both high-fidelity models and experimental data.
  — Development of advanced design optimization tools and approaches.
  — Advancement of site-specific control of microstructure through process parameters for real parts, including complex shapes and complex alloys.
- *Long-term research areas*: Integration of site-specific microstructure control into design optimization.

Several subgroup members also discussed the following nontechnical challenges across AM:

- The lack of stable, long-term research funding;
- A lack of willingness to fund testing and measurement;
- The use of proprietary alloys;
- The lack of a community standard file and standardized formats for experimental and simulation data;

- The need for increased collaboration between domain scientists and computational scientists; and
- The lack of students and staff with necessary expertise such as computational material science, manufacturing, model-based engineering, computer-aided-design-based topology optimized design, and software development for modern computer architecture.

To increase collaboration and better address technical and nontechnical challenges, several subgroup members suggested that industry and academia support efforts that provide foundations for collaboration (e.g., AM-Bench). Industry might consider funding defined challenges in which academia and laboratory teams could compete. Programs could be created for targeted collaborative industry–academia–laboratory research to tackle specific application challenges. These subgroup members emphasized the importance of having adequate, stable funding available over extended time periods and suggested that the U.S. Department of Energy Hubs[1] concept could be applicable for AM. Many subgroup members suggested specific actions that could help address these challenges, including a call for proposals in the industry–academia–laboratory research areas and the expansion of educational programs that are domain specific and multidisciplinary. Some members of this breakout group suggested that a follow-on workshop could address topics such as challenges and opportunities in topology and shape optimization with site-specific microstructure control as well as multidisciplinary educational programs for AM processes.

## MODELING ASPECTS OF PROCESS AND MACHINE DESIGN

*Subgroup Members*

*Mustafa Megahed (ESI Group), Wing Kam Liu (Northwestern University), Jian Cao (Northwestern University), Tahany El-Wardany (United Technologies Research Center), and Winfried Keiper (European Technology Platform for Advanced Engineering Materials and Technologies)*

### Breakout Discussion

Megahed led this breakout group, which discussed modeling aspects of process and machine design. Megahed, Liu, Cao, El-Wardany, and

---

[1] For more information on the U.S. Department of Energy's Hubs, see https://www.energy.gov/science-innovation/innovation/hubs, accessed March 11, 2019.

Keiper proposed the following challenges and research needs for this topic:

- Identifying the source of process variability, which can be done by determining the sensitivity of the process to certain parameters, uncertainty quantification, and process control.
- Calibrating and validating models, even in the absence of experimental data.
- Designing the experiments needed to deliver necessary data.
- Developing a community database for relevant data in standardized forms.
- Advancing models to capture details such as environmental effects, alloying elements, and doping.[2] These may utilize artificial intelligence and machine learning methods.
- Improving the use of data reduction and reduced-order modeling to increase efficiency.

These subgroup members also highlighted three major nontechnical challenges:

- *Data sharing.* The research community would benefit from increased access to data. Several subgroup members speculated that the reticence to share data might be a cultural problem since most researchers are not used to sharing their data. They highlighted nuclear physics databases as a possible example to emulate, particularly the use of a centralized body to help transform raw data into evaluated data. Shared databases also need to be sustainable as well as continually maintained and updated.
- *Interpretable machines.* Manufacturers have historically been reluctant to share the inner workings of their machines for a variety of business reasons. However, these subgroup members noted that having more transparent machine processes would enable research advancements.
- *Interdisciplinary education.* These subgroup members explained that there needs to be a more efficient way of learning about a wide variety of topics relating to AM, including hardware, underlying physics, metrology, algorithm development, optimization, numerical simulation, thermodynamics, statistics, and data analytics.

---

[2] Alloying elements are defined as metallic or nonmetallic elements that are added in specified or standard amounts to a base metal to make an alloy (Business Dictionary, 2019), and doping is the mixing of a small amount of an impurity into a silicon crystal (Brain, 2001).

The subgroup members also discussed partnerships. Several members noted successful models such as America Makes,[3] Horizon 2020,[4] CleanSky,[5] and other data sharing efforts that encourage community databases. Other options could be for industries to enable more internships and fellowships for students and researchers. A number of subgroup members also suggested more partnerships among researchers in the European Union and the United States and among small- and medium-size enterprises; this could encourage more collaboration, data exchange, and international research funding.

For a possible follow-on workshop, some of the subgroup members proposed themes including the definition of joint standards and tolerances, digital twin and threads for AM, interdisciplinary education, and the various intermediate-term challenges and goals that were discussed throughout the workshop.

## ACCELERATING PRODUCT AND PROCESS QUALIFICATION AND CERTIFICATION

*Subgroup Members*

*David Teter (Los Alamos National Laboratory), Jens Telgkamp (Airbus Operations GmbH), Vincent Paquit (Oak Ridge National Laboratory), Paolo Gennaro (GF Precicast Additive SA), Johannes Henrich Schleifenbaum (Fraunhofer Institute for Laser Technology), Richard Ricker (National Institute of Standards and Technology), Josh Sugar (Sandia National Laboratories), and Ben Dutton (Manufacturing Technology Centre)*

### Breakout Discussion

Teter and Telgkamp led the discussion for this subgroup, which focused on accelerating product and process qualification and certification. This discussion was divided into short-term (less than 5 years) and intermediate-term (5 to 10 years) goals that could enable a long-term vision for AM.

Teter explained that the long-term vision is the ability to design, print, and qualify a product correctly the first time. This includes as-built quality, in which people have very limited destructive evaluation for parts

---

[3] For more information on America Makes, see https://www.americamakes.us, accessed March 11, 2019.

[4] For more information on Horizon 2020, see https://ec.europa.eu/programmes/horizon2020/, accessed March 11, 2019.

[5] For more information on CleanSky, see https://www.cleansky.eu, accessed March 11, 2019.

being generated, and built-in quality assurance, in which data are collected as a part is being printed. Modeling and simulation play an important role—a multiphysics process–structure–property–performance prediction is needed. Cybersecurity is another concern, particularly in terms of building resiliency to the threat of fraudulent components over the next 10 years. Several subgroup members noted that the ability to track each part is needed, including attaching the license to build and proof of quality to each part. Lastly, some subgroup members commented on the need for government-to-government agreements on AM with shared objectives, data, and frameworks. They suggested that long-term efforts should focus on the need for AM to be operational and fully accepted by certification groups.

To advance this long-term vision, the subgroup highlighted the following short- and intermediate-term research goals:

- *Short term*: Several subgroup members suggested a short-term focus on AM technology and materials development, such as making the process less sensitive to variability and defects. Below are some specific open challenges that these members highlighted.
  — Improving the understanding of the influence of feedstock parameters, taking into consideration the key material properties and process parameters. These subgroup members emphasized this as a high priority.
  — Developing guidance on sensor technology.
  — Improving the openness of control systems.
  — Refining the definition of "good" data as well as a common test part/object for qualification and microstructure. ASTM F42[6] may be able to help determine goals, objectives of test part/object, and number of object definitions needed.
  — Collecting defect catalogues for critical flaw size and type, frequency, distributions, and criticality of locations. Telgkamp noted that this is particularly important for highlighting research and development needs.
  — Strengthening the understanding of current sensor technology, limits, capabilities, stability, and reliability.
- *Intermediate term*: Several subgroup members suggested an intermediate-term focus on continuous standardization activities, such as development and maturation. Below are some specific open challenges that they identified.

---

[6] For more information on ASTM F42, see https://www.astm.org/COMMITTEE/F42.htm, accessed March 11, 2019.

- Using process monitoring in decision making, such as when and how to repair a part or when to discard it.
- Developing reduced-order models for decision making.
- Maturing tools for sensor data fusion and reduction.
- Exploring machine learning methods to improve microstructure and property predictions.
- Increasing data sharing and establishing a common or global database. These subgroup members noted that this was mentioned throughout the workshop.
- Improving machine-to-machine knowledge transfer.
- Developing high-throughput characterization and development for new and mature sensors, based on the sensing needs to be identified.

## INDIVIDUAL RESPONSE RESULTS

Participants at the workshop were also asked to provide their thoughts on the top priority research needs for advancing AM, top "nontechnical" challenges to commercialization of AM, and actions that could help address these nontechnical challenges. The individual responses were analyzed by a workshop subgroup and summarized by Celia Merzbacher (SRI International). She explained that the technical challenges suggested by the workshop participants centered on needing more AM materials, improving the understanding of microstructure prediction, developing standards and benchmark measurements, and improving in-situ monitoring capability. For nontechnical challenges, she explained that the responses centered on encouraging data sharing, increasing funding, improving training and education, enabling machine transparency, and increasing trust in AM parts. Many participants suggested that these challenges could be approached by increasing coordination and communication among stakeholders, perhaps through more convening activities, collaborations, standards, and funding.

## REFERENCES

Brain, M. 2001. "How Semiconductors Work." April 25. HowStuffWorks.com. https://electronics.howstuffworks.com/diode1.htm, accessed March 11, 2019.

Business Dictionary. 2019. "Alloying Element." http://www.businessdictionary.com/definition/alloying-element.html, accessed March 11, 2019.

# Appendixes

# A

# Registered Workshop Participants

Ranadip Acharya, United Technologies Research Center
Jian Cao, Northwestern University
Michele Chiumenti, Uni Barcelona
Bianca Colosimo, Politecnico di Milano
Laurent D'Alvise, GeonX
Michel Delanaye, GeonX
Shawn Dirk, Sandia National Laboratories
Ben Dutton, Manufacturing Technology Centre
Tahany El-Wardany, United Technologies Research Center
Paolo Gennaro, GF Precicast Additive SA
Jarred Heigel, National Institute of Standards and Technology
Eric Jägle, Max Planck Institute
Kyle Johnson, Sandia National Laboratories
Winfried Keiper, European Technology Platform for Advanced Engineering Materials and Technologies
Jonathan King, Department of Energy
Carolin Körner, Friedrich-Alexander Universität Erlangen-Nürnberg
Deniece Korzekwa, Los Alamos National Laboratory
Christian Leinenbach, Empa
Lyle Levine, National Institute of Standards and Technology
Wing Kam Liu, Northwestern University
Adhish Majmudar, GeonX
Ade Makinde, General Electric Global Research Center

Matthias Markl, Friedrich-Alexander Universität Erlangen-Nürnberg
Mustafa Megahed, ESI Group
Celia Merzbacher, SRI International
Vincent Paquit, Oak Ridge National Laboratory
Janki Patel, National Academies of Sciences, Engineering, and Medicine
Nancy Reid, University of Toronto
Daniel Reznik, Siemens
Richard Ricker, National Institute of Standards and Technology
R. Allen Roach, Sandia National Laboratories
Johannes Schilp, Universität Augsburg Lehrstuhl für Produktionsinformatik
Johannes Henrich Schleifenbaum, Fraunhofer Institute for Laser Technology
Michael Schmidt, Friedrich-Alexander Universität Erlangen-Nürnberg
Michelle Schwalbe, National Academies of Sciences, Engineering, and Medicine
Annett Seide, MTU Aero Engines
Marvin Siewert, University of Bremen
Josh Sugar, Sandia National Laboratories
Amit Surana, United Technologies Research Center
Erik Svedberg, National Academies of Sciences, Engineering, and Medicine
Jens Telgkamp, Airbus Operations GmbH
David Teter, Los Alamos National Laboratory
John Turner, Oak Ridge National Laboratory
Kilian Wasmer, Empa
Shoufeng Yang, KU Leuven

# B

# Workshop Agenda

A Workshop on the Frontiers of Mechanistic Data-Driven Modeling for Additive Manufacturing
October 24–26, 2018

Neue Materialien Fürth GmbH
Dr.-Mack-Straße 81, Technikum 1, 6th Floor
Fürth, Germany

### Day 1: October 24

**9:00 a.m.**     **Welcome from the Co-Chairs**
*Carolin Körner, Co-Chair, Friedrich-Alexander Universität Erlangen-Nürnberg*
*Wing Kam Liu, Co-Chair, Northwestern University*

**9:20 a.m.**     **Opening Comments from the Sponsors**
*R. Allen Roach, Sandia*
*Richard Ricker, NIST*

**9:40 a.m.**     **Opening Comments from the National Academies**
*Michelle Schwalbe, Board on Mathematical Sciences and Analytics*
*Erik Svedberg, National Materials and Manufacturing Board*

## SESSION 1: MEASUREMENTS AND MODELING FOR PROCESS MONITORING AND CONTROL

10:00 a.m. **Introduction to Session 1**
*Bianca Colosimo, Politecnico di Milano*

10:10 a.m. **Measurement Science for Process Monitoring and Control**
*Jarred Heigel, National Institute of Standards and Technology*

10:40 a.m. **Break**

11:00 a.m. **Process Simulation as a Complement of Process Monitoring**[1]
*Daniel Reznik, Siemens*

11:30 a.m. **Lunch**

12:30 p.m. **Panel Discussion**

- Brief introductions and statements of research interests
- Open discussion, led by Bianca Colosimo

*Panelists:*
Bianca Colosimo, Politecnico di Milano
Ben Dutton, Manufacturing Technology Centre
Jarred Heigel, National Institute of Standards and Technology
Daniel Reznik, Siemens
Kilian Wasmer, Empa
Amit Surana, United Technologies Research Center

2:00 p.m. **Break**

## SESSION 2: DEVELOPING MODELS TO REPRESENT MICROSTRUCTURE EVOLUTION, ALLOY DESIGN, AND PART SUITABILITY

2:30 p.m. **Introduction to Session 2**
*Lyle Levine, National Institute of Standards and Technology*

---

[1] Unable to attend.

APPENDIX B

| | |
|---|---|
| 2:35 p.m. | **Measurements for Additive Manufacturing of Metals**<br>*Lyle Levine, National Institute of Standards and Technology* |
| 3:05 p.m. | **Break** |
| 3:30 p.m. | **Predicting Material State and Performance of Additively Manufactured Parts**<br>*Kyle Johnson, Sandia National Laboratories* |
| 4:00 p.m. | **Panel Discussion**<br>• Brief introductions and statements of research interests<br>• Open discussion, led by Lyle Levine<br><br>*Panelists:*<br>*Lyle Levine, National Institute of Standards and Technology*<br>*Eric Jägle, Max Planck Institute*<br>*Kyle Johnson, Sandia National Laboratories*<br>*Christian Leinenbach, Empa*<br>*Deniece Korzekwa, Los Alamos National Laboratory*<br>*Annett Seide, MTU Aero Engines*<br>*John Turner, Oak Ridge National Laboratory* |
| 5:30 p.m. | **Conclude Sessions** |

**Day 2: October 25**

| | |
|---|---|
| 9:00 a.m. | **Recap of Day 1; Major Themes and Overview for the Day**<br>*Session 1: Bianca Colosimo*<br>*Session 2: Lyle Levine* |

**SESSION 3: MODELING ASPECTS OF
PROCESS AND MACHINE DESIGN**

| | |
|---|---|
| 9:30 a.m. | **Introduction to Session 3**<br>*Tahany El-Wardany, United Technologies Research Center* |
| 9:40 a.m. | **Current State of Commercial Powder-Bed Additive Machines—Improvements Needed to Minimize Build-to-Build Variability**<br>*Ade Makinde, General Electric Global Research Center, with support from Johannes Henrich Schleifenbaum, Fraunhofer Institute for Laser Technology, and Shoufeng Yang, KU Leuven* |

10:10 a.m.   Break

10:30 a.m.   **Modeling Challenges and Opportunities at the Part Level**
*Jian Cao, Northwestern University, with support from Ranadip Acharya, United Technologies Research Center, and Mustafa Megahed, ESI Group*

11:00 a.m.   **Panel Discussion**
- Brief introductions and statements of research interests
- Open discussion, led by Tahany El-Wardany

<u>Panelists</u>:
*Tahany El-Wardany, United Technologies Research Center
Ranadip Acharya, United Technologies Research Center
Jian Cao, Northwestern University
Ade Makinde, General Electric Global Research Center
Mustafa Megahed, ESI Group
Johannes Henrich Schleifenbaum, Fraunhofer Institute for Laser Technology
Michael Schmidt, Friedrich-Alexander Universität Erlangen-Nürnberg
Shoufeng Yang, KU Leuven*

12:30 p.m.   Lunch

## SESSION 4: ACCELERATING PRODUCT AND PROCESS QUALIFICATION AND CERTIFICATION

1:30 p.m.    **Introduction to Session 4**
*Paolo Gennaro, GF Precicast Additive SA*

1:35 p.m.    **Process Qualification and Technological Validation, from Casting to Additive**
*Paolo Gennaro, GF Precicast Additive SA*

2:05 p.m.    **Modeling and Simulation**
*Michel Delanaye,[2] GeonX*

2:35 p.m.    Break

---

[2] Unable to attend.

APPENDIX B

| | |
|---|---|
| 3:00 p.m. | **Panel Discussion**<br>• Brief introductions and statements of research interests<br>• Open discussion, led by Paolo Gennaro<br><br>*Panelists:*<br>Paolo Gennaro, GF Precicast Additive SA<br>Vincent Paquit, Oak Ridge National Laboratory<br>Jens Telgkamp, Airbus Operations GmbH<br>Michel Delanaye, GeonX<br>David Teter, Los Alamos National Laboratory<br>Richard Ricker, National Institute of Standards and Technology |
| 5:00 p.m. | **Conclude Presentations and Discussions** |

**Day 3: October 26**

| | |
|---|---|
| 9:00 a.m. | **Recap of Day 2; Major Themes and Overview for the Day**<br>*Session 3: Tahany El-Wardany*<br>*Session 4: Paolo Gennaro* |
| 9:30 a.m. | **Breakout Groups**<br>• Measurements and Modeling for Process Monitoring and Control<br>• Developing Models to Represent Microstructure Evolution, Alloy Design, and Part Suitability<br>• Modeling Aspects of Process and Machine Design<br>• Accelerating Product and Process Qualification and Certification |
| 12:00 p.m. | **Lunch (with Breakout Groups)** |
| 1:00 p.m. | **Breakout Groups Report Back** |
| 2:30 p.m. | **Final Comments from Co-Chairs and Sponsors**<br>*Carolin Körner, Co-Chair, Friedrich-Alexander Universität Erlangen-Nürnberg*<br>*Wing Kam Liu, Co-Chair, Northwestern University*<br>*Sponsors' Representatives* |
| 3:00 p.m. | **Adjourn Workshop** |

# C

# Workshop Statement of Task

A National Academies of Sciences, Engineering, and Medicine-appointed ad hoc committee will plan and organize a 3-day workshop to explore the frontiers of integrated data-driven modeling for additive manufacturing. This workshop will convene leading experts in online monitoring, science of materials and mechanics, optimization and controls, and qualification and certification from the United States and the European Union to discuss approaches to and challenges with the following:

- Measuring and modeling process monitoring and control;
- Developing models to represent microstructure evolution, alloy design, and part suitability;
- Modeling phases of process and machine design; and
- Accelerating product and process qualification and certification.